ALLE ZEIT WACH
1842

S. Suga · N. Watabe (Eds.)

Hard Tissue Mineralization and Demineralization

With 106 Figures

Springer-Verlag
Tokyo Berlin Heidelberg New York London
Paris Hong Kong Barcelona Budapest

Shoichi Suga, d.d.s., d.m.s.
Professor, Department of Pathology,
The Nippon Dental University School of Dentistry at Tokyo,
1-9-20 Fujimi, Chiyoda-ku, Tokyo, 102 Japan

Norimitsu Watabe, d.sc.
Director of Electron Microscopy Center,
Professor of Biological Sciences and Marine Science Program,
University of South Carolina,
Columbia, SC 29208, USA

ISBN-13:978-4-431-68185-4 e-ISBN-13:978-4-431-68183-0
DOI: 10.1007/978-4-431-68183-0

Library of Congress Cataloging-in-Publication Data. Hard tissue mineralization and demineralization/S. Suga, N. Watabe (eds.). p. cm. "Papers mainly presented at the open seminar held in Tokyo, October 13, 1990, under the auspices of the Society of the Nippon Dental University." Includes bibliographical references and index. ISBN-13:978-4-431-68185-4 1. Biomineralization—Congresses. I. Suga, S. (Shōichi), 1927– . II. Watabe, Norimitsu, 1922–
QH512.H37 1992. 591.19′214—dc20. 92-17466

Softcover reprint of the hardcover 1st edition 1992

Preface

Mineral formation occurs commonly in many organisms ranging from bacteria to humans within and/or outside the living tissues. More than forty types of such biominerals have been reported to date, and two of the most common are calcium carbonates and calcium phosphates. Much work has been carried out recently to elucidate the mechanisms of biomineralization. The studies are concentrated in the areas of calcium phosphate mineral formation in vertebrate bone and teeth and calcium carbonate mineral formation in coral skeletons and molluscan shells. Demineralization of bone and teeth has also been extensively investigated as it relates to human health, and this phenomenon has become recognized also in invertebrate skeletal systems. The discussions presented here are also concerned with those areas. We hope that this volume will stimulate interactions and cooperation between the investigators of invertebrate and vertebrate biomineralization to increase our understanding of the mechanisms of biological mineral formation and demineralization.

This volume is a collection of papers mainly presented at the open seminar held in Tokyo, October 13, 1990, under the auspices of The Society of The Nippon Dental University. The objective of the seminar was to introduce the status of biomineralization research to investigators in many different fields who had not been able to participate in the Odawara Symposium (the Sixth International Symposium on Biomineralization, October 9–12, 1990).

We are grateful to The Society of The Nippon Dental University and Lion Corporation, Japan whose generous financial support made the open seminar and publication of this volume possible. We also thank the staff of Springer-Verlag Tokyo for their constant assistance in the editing and production of this volume.

SHOICHI SUGA
NORIMITSU WATABE

Contents

List of Contributors

Towards an Understanding of Tissue-Specific Biomineralization

Harold C. Slavkin[1], Ching C. Hu[1], Yasuyuki Sasano[2], Sasson David[1], Pablo Bringas, Jr.[1], Yasunori Sakakura[1], Valentino Santos[1], and Masanori Nakamura[2]

Summary. We propose to test the hypothesis that time and position-restricted growth factor expression regulates dentine and enamel biomineralization. We predict that a developmental program which defines timing and position for the putative dentine-enamel junction (pDEJ) serves to initiate the nucleation process which propagates either as (1) dentine mineral deposition, or (2) enamel mineral deposition patterns. A simple organ culture system, devoid of exogenous serum or plasma supplementation, provides a controlled model to test our hypothesis and these predictions. Dentine- and enamel-specific biomineralization were observed when E15 mouse cap stage tooth organs were cultured for 3 weeks, and root and alveolar bone formations were observed within 4 weeks using serumless, chemically defined medium. Under these permissive conditions, a developmental program expressed the DEJ and biomineralization. The initial DEJ constituents appeared to provide instructions for nucleation and specificity for dentine and enamel crystal size, orientation, number, and growth patterns. In tandem, EGF and EGF receptors were expressed by the same cells in this model system, suggesting autocrine regulation. These results have provided a rationale towards understanding tissue-specific biomineralization.

Key words: Enamel—Dentine—Grwoth factors—In vitro—Mouse

Introduction

Cells and their extracellular matrices (ECM) control tissue-specific biomineralization [1–2]. Anionic macromolecules (e.g., phospholipids and phosphoproteins) are associated with biomineralization throughout vertebrate evolution and are implicated in the control processes for biomineralization [1–4] (Fig. 1). However, the determinants for the timing, position, and unique specificities

[1] Center for Craniofacial Molecular Biology, School of Dentistry, University of Southern California, Los Angeles, CA 90033, USA
[2] Department of Anatomy, School of Dentistry, Tohoku University, Sendai, 980 Japan

Cell- and extracellular matrix-mediated biomineralization

Secretion of extracellular matrix constituents

↓

Intra- and intermolecular interactions

↓

Induction and propagation of tissue-specific biomineralization (e.g., position, size, and patterns of crystal morphogenesis)

Fig. 1. Cell- and extracellular matrix-mediated biomineralization

for structural gene products associated with biomineralization are not as yet known.

Calcium phosphate formation resulting in calcium hydroxyapatite crystal (HAP) morphogenesis in bone, dentine, cementum, and enamel are associated with heterogeneous nucleation processes, controlled by specific regulatory and structural gene products within temporally and positionally restricted extracellular microenvironments. It is assumed that ECM organic components (e.g., type I collagen, phosphoproteins, phospholipids, etc.) function as nucleators or substrates [1–4]; chemical groups (e.g., amino acid side chains, glycosylations, and/or phosphorylations in the case of proteins) can produce intermolecular three-dimensional stereochemical geometry with attendant electrical charge distribution which replicates specific crystal planes of the HAP crystal to be nucleated [1–2]. The nucleation substrate for HAP crystal formation is assumed to be a solid state aggregate. Therefore, the specific physicochemical properties of the solid state aggregate, such as structural details characteristic for each of the dental tissue ECM proteins, provides tissue-specific biomineralization (e.g., number, size, and patterns of crystal growth). How these solid aggregates are formed and how they induce and control tissue-specific patterns of HAP crystal growth are not known.

The mouse tooth organ provides an excellent model for a number of questions related to the timing, position, and sequential inductive processes which result in cellular differentiation to become specific phenotypes related to biomineralization [5–11]. Enamel and dentine provide two curiously different and unique tissue examples of ECM protein-mediated biomineralization. Ectodermally derived epithelia synthesize and secrete enamel proteins to form an ECM in juxtaposition to and coordinated with ectomesenchymally derived odontoblasts which synthesize and secrete dentine proteins to form and dentine ECM [3, 8]. At the interface between dental inner enamel epithelia and adjacent odontoblasts, hole zone regions formed within type I collagen fibrils, as well as matrix vesicles, may provide *de novo* nucleation sites for initial HAP crystal formation [4, 12]. These initial processes result in the position of the putative dentine-enamel junction (pDEJ) [12]. From the initial DEJ, relatively small and randomly arranged HAP crystals increase in number (multiplication) but not in size, and propagate dentine-specific biomineralization. In juxtaposition, HAP crystals from the DEJ grow in length along their *c*-axis and become highly ordered, enamel-specific biomineralization. Two different tissue-specific patterns of biomineralization appear to be initiated at the same interface (DEJ), yet grow to different sizes and patterns in opposite directions, presumably regulated by two very different cell/tissue phenotypes. Whereas many ultrastructural and physicochemical details of this process have been described [reviews 1–4], tissue-specific mechanisms remain to be elucidated.

Recently, dentine and enamel tissue-specific biomineralization were demonstrated in a simple in vitro model without endocrinic or long-range humoral supplements. Embryonic mouse molar tooth organs (E15/E16, early cap stage) expressed dentine- and enamel-specific patterns of HAP crystal biomineralization in a serumless, chemically defined medium [5–6]. When these cultures were extended for periods up to 31 days in vitro, alveolar bone was also expressed in association with the forming tooth organ roots [7]. The culture medium was devoid of transferrin, serum, or other exogenous growth/regulatory factors (e.g., epidermal growth factor, transforming growth factors alpha or beta, nerve growth factor, insulin-like growth factors, vitamin A, vitamin D, etc.). Therefore, endocrine-type, long-range regulatory molecules (e.g., steroid or polypeptide hormones, vitamin D, serum or plasma factors, etc.) were not pre-requisite for dentine- or enamel-specific biomineralization.

The present contribution summarizes and reviews the results of these experimental strategies and attempts to identify potential rules for biomineralization. We anticipate that multiple regulatory factors determine when, where, and how tissue-specific biomineralization takes place. It is anticipated that the suggested criteria might serve to further catalyze efforts towards defining the mechanisms which regulate the determinants for biomineralization.

Material and Methods

Embryonic Mouse Molar Dissections and a Simple Organ Culture System

Embryonic mandibular first molars (M_1) were dissected from 15-days gestation (E15), timed-pregnant Swiss Webster strain mice and then cultured for periods up to 31 days in vitro using a serumless, chemically defined medium, as previously described [5–7].

Von Kossa Histochemistry

Non-demineralized cultured explants and control in vivo specimens (birth, days 2 and 7 postnatal) were processed for histochemistry to determine the deposition of calcium phosphate salts, as previously described [5–7].

Anhydrous Fixation, Transmission Electron Microscopy, and Electron Diffraction Studies

Non-demineralized specimens were processed by the Landis method as previously reported [13] and physical studies as previously described [6].

Immunocytochemistry

Anti-amelogenin antibody and antibodies designed to identify motifs (-LPPHPGHPGYI-) within the amelogenin primary structure were used for immunohistochemistry by methods previously described [14, 15]. In addition, polyclonal anti-EGF antibodies were used to identify the translation product within odontogenic cells.

In Situ Hybridization

AMEL (amelogenin) transcripts were identified and localized using ^{35}S-labeled sense and antisense RNA probes by methods previously reported [16, 17].

EGF mRNA Phenotyping

The RT-PCR strategy (reverse transcriptase, polymerase chain reaction) for mRNA phenotyping, as recently described [18–20], was used to detect relatively low levels of EGF transcripts during in vitro tooth development.

Results

Summary of Tooth Development In Vivo

The salient descriptions of mouse tooth organ development are readily found in a number of very useful reviews [e.g., 8–11]. Figure 2 outlines the sequence and timing of major stages in mouse mandibular first molar tooth organ (M_1) development. Dentine phosphoprotein was identified within odontoblasts prior to biomineralization, but was not secreted until after the DEJ was formed. Initial biomineralization was defined using ultrastructural criteria for HAP crystal formation associated with matrix vesicles and collagen. In M_1, this was observed along the major buccal cusp 24 h prior to birth. Anionic enamel protein(s) represented 46 kd (pI, 5.5) and 72 kd (pI, 5.8) proteins which were first expressed prior to AMEL and shared one or more epitopes with enamel proteins.

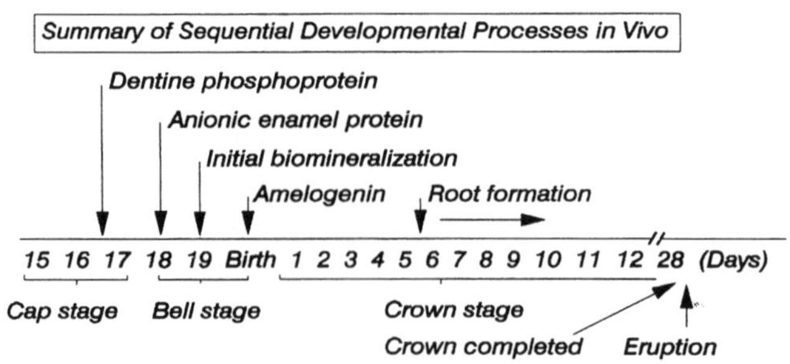

Fig. 2. Summary of sequential developmental processes in vivo

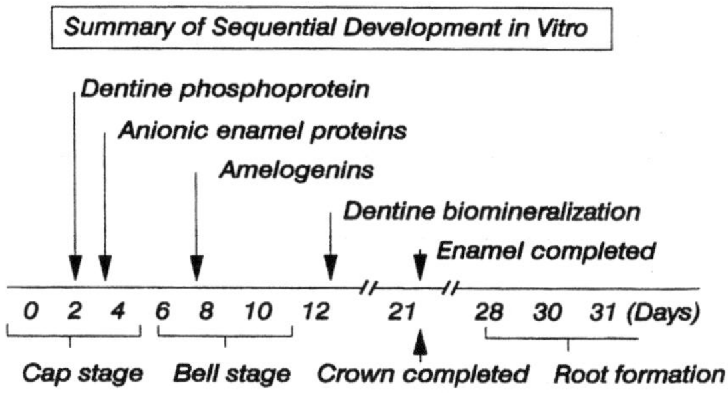

Fig. 3. Summary of sequential developmental processes in vitro

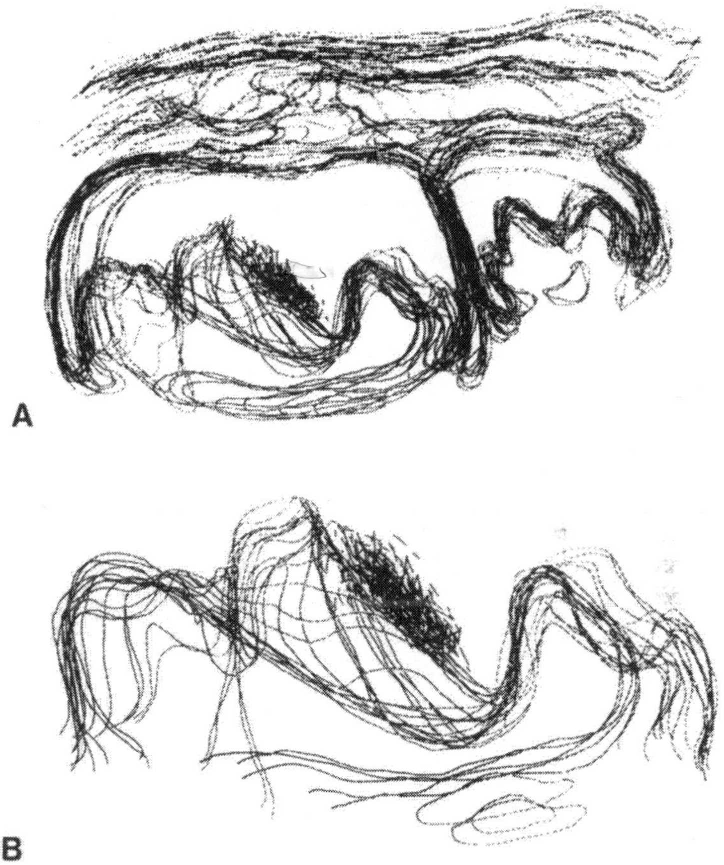

Fig. 4. Computer-assisted, three-dimensional reconstructions of AMEL localization using in situ hybridization. **A** AMEL transcripts were first observed along the distal surface of the major buccal cusp in M_1 at 6 days in vitro. The E15 cap stage explant with oral epidermis and enamel organ epithelia produced both M_1 and M_2 molars. **B** AMEL transcript localization patterns were comparable to those described for in vivo development

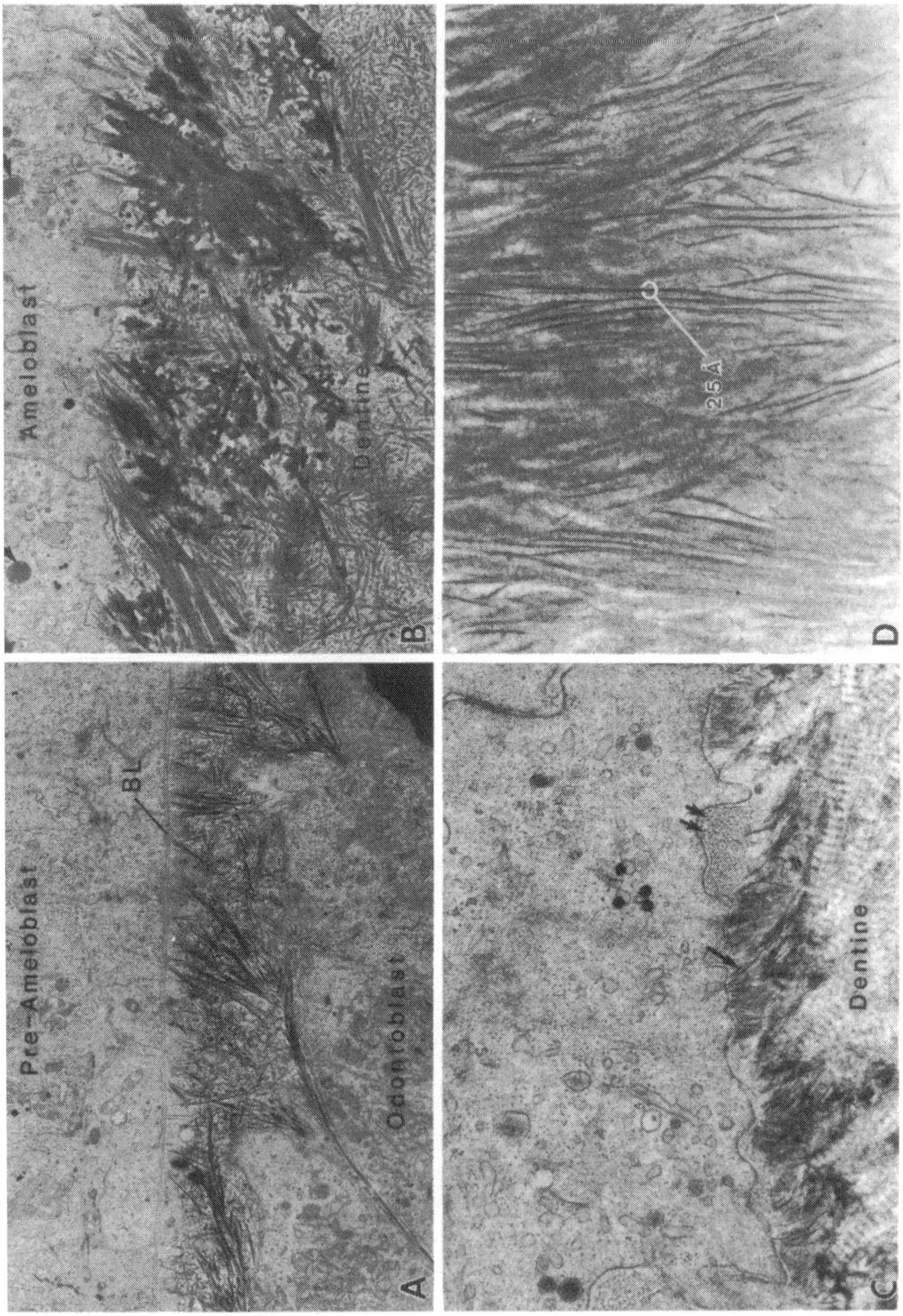
Ameloblast
Dentine
B
25Å
D
Pre-Ameloblast
BL
Odontoblast
A
Dentine
C

Fig. 5. Ultrastructural survey of dentine-enamel junction (DEJ) formation and tissue-specific biomineralization. **A** Preameloblasts with a continuous basal lamina (*BL*) expressed anionic enamel proteins prior to biomineralization and AMEL expression. Odontoblasts produced matrix vesicles, type I collagen and dentine phosphoprotein prior to biomineralization. **B** In tandem with basal lamina degradation and removal from ameloblasts, biomineralization along the putative DEJ appeared as disparate aggregates of HAP crystals associated with either matrix vesicles or type I collagen. AMEL translation products were packaged within secretory vesicles (*arrows*). **C** Enamel crystals (*arrow*) appeared to form at the DEJ. Electron-dense, granular materials (AMEL) were secreted from ameloblasts beginning at birth (*double arrows*). **D** Enamel crystals were approximately 25 Å in diameter

Summary of Tooth Development In Vitro

The embryonic mouse molar tooth organ has been used in a number of in vitro studies [reviews, 8–11]. Figure 3 outlines the sequence and timing of M_1 development using serumless, chemically defined medium [5–7]. Albeit slower and smaller in scale, the sequence of the observed tooth development was comparable to the sequence observed for in vivo development.

The Sequence and Timing of Tissue-Specific Biomineralization

The major buccal cusp of the M_1 region demonstrated the initial ultrastructural evidence for tissue-specific biomineralization. A detailed description of this process has recently been reported by our laboratory [14, 16–17]. HAP crystals were first observed within matrix vesicles in the position of the future DEJ region, in juxtaposition to inner enamel epithelia with continuous basal lamina and adjacent odontoblast cell processes. Small HAP crystals were observed in association with type I collagen fibrils. HAP crystals formed aggregates, the aggregates coalesced and subsequently produced the mineralizing dentine ECM.

In the M_1 model, initial biomineralization was indicated approximately 24 h before birth. The timing and position of AMEL gene expression in vivo and in vitro were determined using in situ hybridization. Inner enamel epithelia initially expressed amelogenin (AMEL) transcription and translation at birth [16], *after* initial biomineralization [14]. When M_1 E15/16 explants were cultured, AMEL was detected at 6 days along the distal buccal surface of the major buccal cusp (Fig. 4A,B). Subsequent patterns for AMEL expression were comparable to those observed for in vivo amelogenesis [16–17].

Figure 5 provides an ultrastructural survey of the formation of the DEJ. A continuous basal lamina was characteristic of the post-mitotic preameloblasts prior to initial biomineralization (Fig. 5A). Preameloblasts expressed non-collagenous, presumably anionic enamel proteins prior to AMEL expression at birth. The basal lamina was removed from the preameloblasts coincident with the formation of the DEJ and biomineralization (Fig. 5B). AMEL was first expressed at birth and characteristic enamel-specific HAP crystal patterns were detected (Fig. 5C).

Anhydrous fixation methods without subsequent staining provided a methodology to observe electron-dense images representing the inorganic constituents of forming enamel and dentine (Fig. 6). Our observations of the DEJ were interpreted to suggest that enamel HAP crystals appeared to grow from the dentine collagen HAP crystals. These observations confirmed those previously reported by Arsenault and Robinson [12].

Endogenous EGF Expression

Recently, Snead and colleagues reported that EGF transcripts were identified in E14 embryonic mouse odontogenic tissues [21]. Earlier, Adamson et al. reported that EGF receptors (EGFR) were first identified at 9-days gestation in

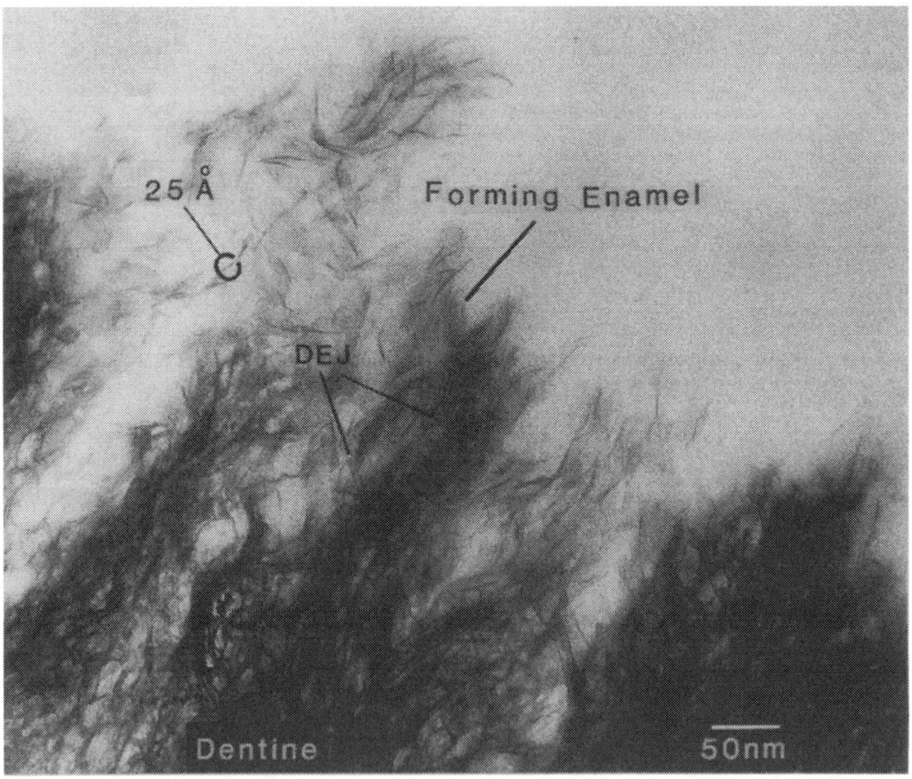

Fig. 6. Enamel HAP crystals appeared to form from dentine HAP crystals along the DEJ when observed in M_1 specimens processed by anhydrous fixation and without staining

Fig. 7. EGF mRNA phenotyping during M_1 development in vivo and in vitro. *PN*, Postnatal development selected was at birth

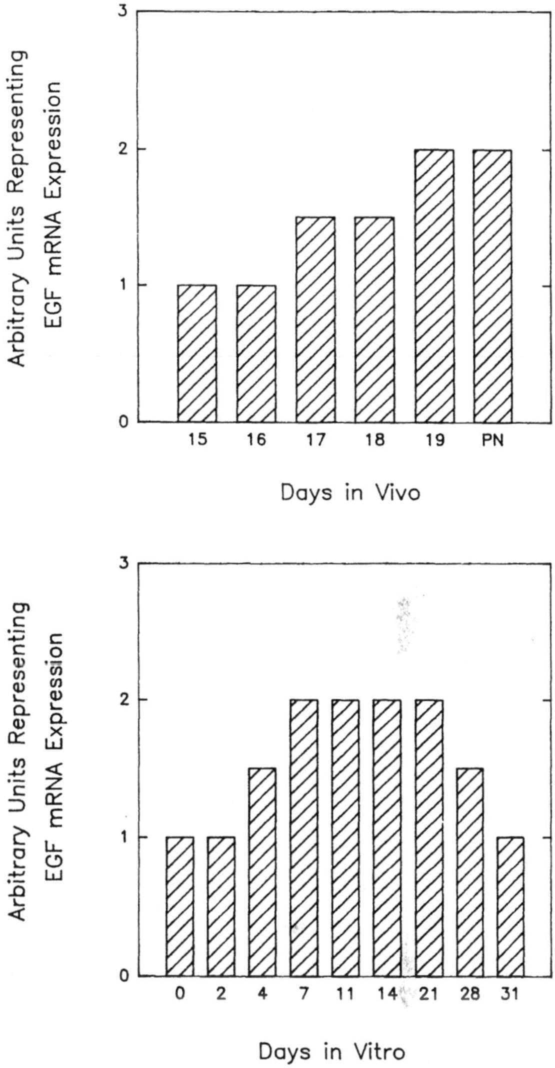

mouse embryos [22]; EGFR were also identified on early embryonic mouse dental tissues [23].

These results were used to formulate our hypothesis that endogenous EGF, and possibly other growth factors, function as either autocrine and/or paracrine regulatory molecules for biomineralization during tooth development. More recent studies in our laboratory demonstrated that EGF transcripts and translation products were expressed during tooth development: the relative levels and patterns of expression changed during development [20, 24]. Figure 5 provides a survey of relative EGF mRNA levels of expression during tooth organ development in vivo and in vitro. A two-fold increase in EGF transcription was observed between the cap and bell stages (e.g., dentine and enamel biomineralization). Figure 6 summarizes the distribution of EGF translation product and EGFR during tooth development. Immunocytochemical evidence showed that cells which produced EGF also contained EGFR (Fig. 7), suggesting that these cells may function by autocrinic regulation [24].

Complementary studies using tyrphostin (10–60 μM) inhibited signal transduction mediated by EGF binding to EGFR with subsequent autophosphorylation of tyrosine residues, produced significant reductions in DNA synthesis, and retarded or inhibited odontogenic cell/tissue differentiations [24]. E15/16 M_1 responded to exogenous EGF (2–20 ng/ml) by a dose-response increase in DNA synthesis [24], whereas E17–E19 M_1 explants responded by increased total protein synthesis and selective inhibition of type I collagen and enamel proteins [25]. In tandem, preliminary studies found that M_1 explants cultured in serumless medium also expressed TGF-alpha, TGF-beta$_1$, IGF-I, IGF-II, bFGF and a number of other putative regulatory factors possibly associated with tissue-specific biomineralization.

Discussion

This analysis provides evidence in support of three potentially important predictions related to tissue-specific biomineralization:

1. A genetically-defined developmental program controls morphogenesis, differentiation, and tissue-specific biomineralization.
2. A series of sequential epithelial-mesenchymal interactions, mediated through autocrine and/or paracrine controls, regulates the timing and position for the future DEJ.
3. The putative DEJ is producted by both epithelial- and mesenchymal-derived ECM products. These products serve as nucleation mediators for HAP crystal formation, and tissue-specific microenvironments are either permissive for multiplication of small and randomly arranged HAP crystals as in dentine, or for increased growth along the *c*-axis resulting in large, ordered HAP crystals associated with enamel.

The following is selected evidence which further substantiates the rationale for these assertions.

Developmental Genetic Program For Tissue-Specific Biomineralization

Regulatory and/or structural genes and their products required for biomineralization have not yet been identified. Whereas ultramicromethods have provided detection for relatively few transcripts and their translation products in embryonic and fetal mouse dental tissues, direct evidence implicating one or more regulatory factors with specific processes related to biomineralization is not yet available.

However, three lines of evidence should be considered towards understanding the genetic program for tissue-specific biomineralization. First, a recent report has demonstrated that a homeotic gene transcript, (i.e., Hox 7.1, a mouse homologue for the *Drosophila* bicoid homebox), was uniquely expressed and coordinated with each mouse tooth organ during the cap stage of development in vivo as well as in vitro [27]. Second, Hox 7.1 expression appeared to be coordinated with endogenous growth factor and receptor expressions prior to and during dental tissue-specific biomineralization (e.g., EGF, TGF-alpha, TGF-beta$_1$, IGF-I, IGF-II, bFGF) [20–21, 23–24, 26]. Third, differential expression of substrate adhesion molecules (e.g., tenascin/cytotactin and syndecan) as well as ECM constituents (e.g., fibronectin and types I and III collagens) were also coordinated with early to late cap stages of tooth development [reviews 8–11, 23, 28]. These three complementary lines of evidence suggest that coordinated and hierarcal gene regulation provides the controls for the timing and position of tissue-specific biomineralization. We suggest that a sequence of multiple and coordinated genetically controlled processes are required to enable cell- and/or ECM-mediated biomineralization.

Highly specific inhibition or perturbation studies, using either antisense or immunoperturbation strategies, might be useful to correlate a specific growth factor, or pattern of growth factor expressions, with a specific developmental process related to biomineralization. The experimental strategy of constructing the so-called "knock-out" or deletion experiments to prevent gene expression within a transgenic mouse model system could significantly enhance an understanding of structure/function during tissue-specific biomineralization.

Criteria Towards Understanding the Molecular Biology of Tissue-Specific Biomineralization

International agreement for criteria is essential towards defining when, where, what, and how cell- and/or ECM-derived constituents regulate tissue-specific biomineralization. The following criteria are suggested for studies designed to determine genes controlling biomineralization: (1) Gene(s) must be expressed during development at precise times and positions concurrent with initiation and propagation of tissue-specific biomineralization (e.g. enamel, dentine, bone, and cementum); and (2) Gene mutations must produce abnormalities in biomineralization (e.g., amelogenesis imperfecta, dentinogenesis imperfecta, and osteogenesis imperfecta).

DEJ Is Formed by both Epithelia and Mesenchyme-Derived Products Which Control Tissue-Specific Patterns of Biomineralization

Prior to and during the formation of the ECM interface, interposed between inner enamel epithelia and adjacent preodontoblast/odontoblasts (e.g., late

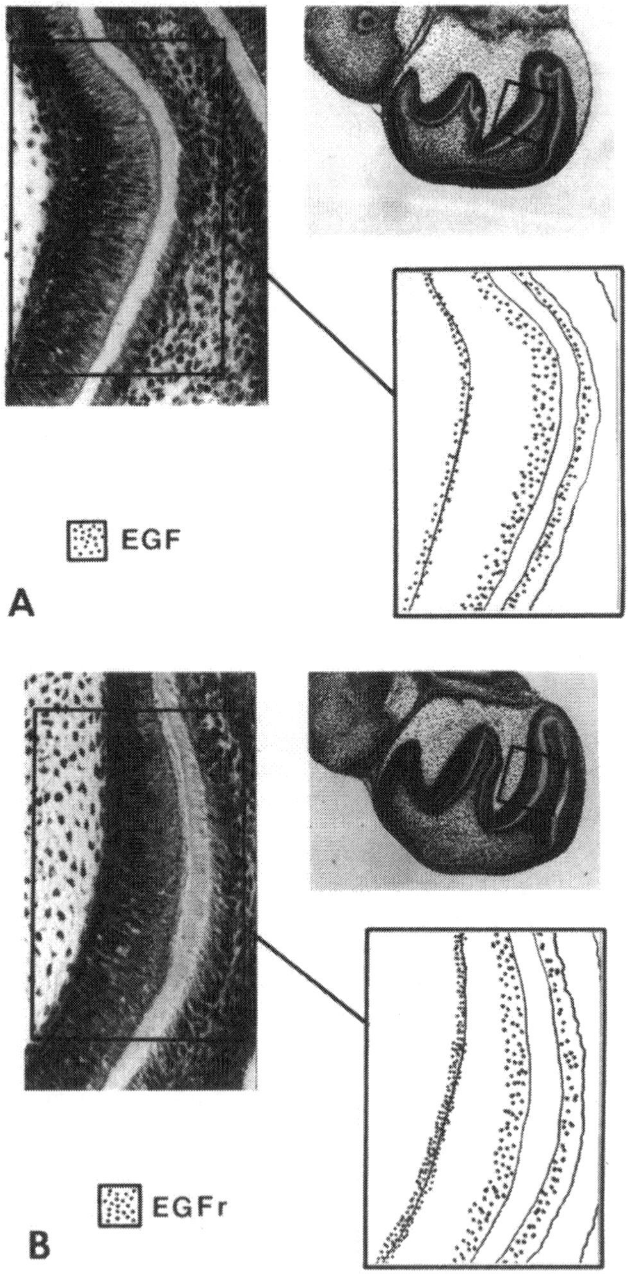

Fig. 8. *EGF* presumably functions in an autocrine mechanism during M_1 development. **A** Immunolocalization of EGF during cap stage development. The region within the rectangle indicating the mesial buccal cusp of the M_1 is shown at low magnification, enlargement, and schematically with immunostained cells indicated by *stippled material*. **B** Immunolocalization of *EGFR* (receptor) in comparable specimens as shown in **A**

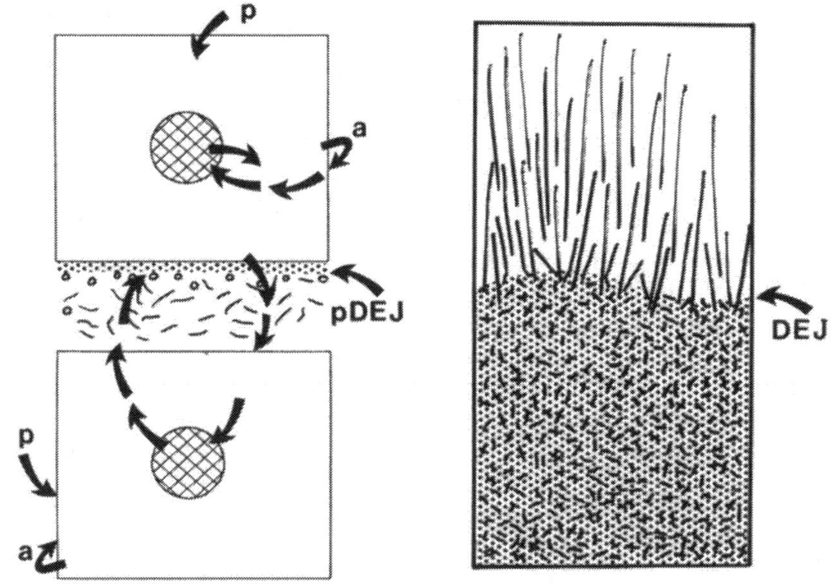

Fig. 9. Evidence supports the hypothesis that both autocrine (*a*) and/or paracrine (*p*) mechanisms possibly control formation of the putative dentine-enamel junction (*pDEJ*). The pDEJ appeared to provide nucleation for HAP crystal formation which proceeds as either (1) small, randomly arranged crystals in dentine, or (2) large, highly ordered HAP crystals in enamel

cap stage of tooth development), both tissue types synthesize and secrete regulatory factors as well as macromolecular products into the forming ECM interface (Fig. 8). Growth factors (e.g., EGF, TGF-beta$_1$, IGF-I, and IGF-II) were sequentially produced and regulated developmental processes through either autocrine or paracrine mechanisms. ECM products include preodontoblast-derived matrix vesicles, types I and III collagens, proteoglycans, substrate adhesion molecules, sequestered growth factors, and epithelial-derived anionic enamel proteins [14]. We suggest that the initial nucleation of HAP crystals within matrix vesicles and that associated with collagen fibrils along the putative DEJ were the resultants of a microenvironment produced and propagated by both epithelial and mesenchymal tissues, and instructive for biomineralization. We further suggest that the putative DEJ, a macromolecular aggregate precipitated at the interface between preameloblasts and odontoblasts, functions as the nucleation site for both enamel and dentine, as previously suggested by Arsenault and Robinson [12]. We assume that subsequent AMEL expression *following the initiation of biomineralization* may function to regulate the size, pattern, and direction of enamel HAP crystal growth.

Acknowledgements. Often in science a hypothesis and supporting evidence which can facilitate the realization and significance of the original hypothesis is published well in advance of other discoveries. An example for this is found in the original hypothesis, which predicted that procaryotic as well as eucaryotic cells interpret and function within their microenvironments through a metabolic code. This idea has been essential for our recent research studies. Therefore, the authors wish to especially acknowledge the intellectual legacy provided through the original hypothesis for a metabolic code put forth by the late Professor Gordon Tomkins (The Metabolic Code. *Science* 189: 760–768, 1975) preceding the discoveries of cytokines, growth factors, and autocrine or paracrine regulation during development. These studies were supported in part by research funds granted by the NIH, USPHS (DE-02848, DE-06425).

References

1. Lowenstam HA, Weiner S (1989) On biomineralization. Oxford University Press, New York, pp 1–324
2. Mann S, Webb J, Williams RJP (1989) Biomineralization. VCH, New York, pp 1–541
3. Veis A (1989) Biochemical studies of vertebrate tooth mineralization. In: Mann S, Webb J, Williams RJP (eds) Biomineralization. VCH, New York, pp 189–222
4. Glimcher MJ (1989) Mechanism of calcification: Role of collagen fibrils and collagen-phosphoprotein complexes in vitro and in vivo. Anat Rec 224(2):139–153
5. Bringas P, Nakamura M, Nakamura E, Evans JC, Slavkin HC (1987) Ultrastructural analysis of enamel formation during in vitro development using chemically-defined medium. Scan Electron Microsc 1:1103–1108
6. Evans J, Bringas P, Nakamura M, Nakamura E, Santos V, Slavkin HC (1988) Metabolic expression of intrinsic developmental programs for dentine and enamel biomineralization in serumless, chemically-defined, organotypic culture. Calcif Tissue Int 42:1–12
7. Slavkin HC, Bringas P, Bessem C, Santos V, Nakamura M, Hsu M-Y, Snead ML, Zeichner-David M, Fincham AG (1988) Hertwig's epithelial root sheath differentiation and initial cementum and bone formation during long-term organ culture of mouse mandibular first molars using serumless, chemically-defined medium. J Periodont Res 23:28–40
8. Slavkin HC (1990) Molecular determinants of tooth development: A review. Crit Rev Oral Biol 1:1–16
9. Thesleff I, Hurmerinta K (1981) Tissue interactions in tooth development. Differentiation 18:75–88
10. Kollar EJ (1983) Epithelial-mesenchymal interaction in the mammalian integument. In: Sawyer RH, Fallon FJ (eds) Epithelial-mesenchymal interaction in development. Praeger, New York, pp 27–50
11. Lumsden AGS (1988) Spatial organization of the epithelium and the role of neural crest cells in the initiation of the mammalian tooth germ. Development 103:155–169
12. Arsenault LR, Robinson BW (1989) The dentino-enamel junction: A structural and microanalytical study of early mineralization. Calcif Tissue Int 45:111–121

13. Landis WJ (1983) Special techniques for problem tissues. In: Jones BR (ed) Electron microscopy. Academic, New York, pp 459–476
14. Slavkin HC, Bessem C, Bringas P, Zeichner-David M, Nanci A, Snead ML (1988) Sequential expression and differential function of multiple enamel proteins during fetal, neonatal and early postnatal stages of mouse molar organogenesis. Differentiation 37:26–39
15. Nanci A, Bendayan M, Slavkin HC (1985) Enamel protein biosynthesis and secretion in mouse incisor secretory ameloblasts as revealed by high resolution immunocytochemistry. J Histochem Cytochem 33:1153–1160
16. Snead ML, Luo W, Lau EC, Slavkin HC (1988) Spatial- and temporal-restricted pattern for amelogenin gene expression during mouse molar tooth organogenesis. Development 104:77–85
17. David S, Bringas P, Santos V, Luo W, Slavkin HC (1991) Individual molar cusp units express independent amelogenin (AMEL) transcription. J Dent Res 70:271a
18. Rappolee DA, Mark D, Banda MJ, Werb Z (1988) Wound macrophages express TGF-alpha and other growth factors in vivo: Analysis by mRNA phenotyping. Science 241:708–712
19. Rappolee DA, Brenner CA, Schultz R, Mark D, Werb Z (1988) Developmental expression of PDGF, TGF-alpha and TGF-beta genes in preimplantation mouse embryos. Science 241:1823–1826
20. Slavkin HC, Snead ML, Luo W, Bringas P, Kikunaga S, Sasano Y, Bessem C, Mayo M, MacDougall M, Rall LB, Rappolee D, Werb Z (1990) Analysis of embryonic cartilage and bone induction in a defined culture system. In: Pecile A, de Bernard B (eds) Bone regulatory factors. Plenum, New York, pp 55–68
21. Snead ML, Luo W, Oliver P, Nakamura M, Don-Wheeler G, Bessem C, Slavkin HC (1989) Localization of epidermal growth factor precursor in tooth and lung during embryonic mouse development. Dev Biol 134:420–429
22. Adamson ED, Deller MJ, Warshaw JB (1981) Functional EGF receptors are present on mouse embryo tissues. Nature 291:656–659
23. Partanen AM, Thesleff I (1989) Growth factors and tooth development. Int J Dev Biol 33:165–172
24. Hu CC, Slavkin (1991) Mitogenic effect of exogenous EGF on developing mouse tooth buds. J Dent Res 70:270a
25. Hata R, Bessem C, Bringas P, Hsu M-Y, Slavkin HC (1990) Epidermal growth factor regulates gene expression of both epithelial and mesenchymal cells in mouse molar tooth organs in culture. Cell Biol Int Rep 14:509–519
26. D'Souza RN, Happonen RP, Ritter NM, Butler WT (1990) Temporal and spatial patterns of transforming growth factor-beta$_1$ in developing rat molars. Arch Oral Biol 35:957–965
27. Mackenzie A, Leeming GL, Jowett AK, Ferguson MWJ, Sharpe PT (1991) The homeobox gene Hox 7.1 has specific regional and temporal expression patterns during early murine craniofacial embryogenesis, especially tooth development in vivo and in vitro. Development 111:269–285
28. Slavkin HC, MacDougall M, Zeichner-David M, Oliver P, Nakamura M, Snead ML (1988) Molecular determinants of cranial neural crest-derived odontogenic ectomesenchyme during dentinogenesis. Am J Med Genet 4:7–22

Fluoride and Iron Concentrations in the Teeth of Primitive Bony Fishes (Osteichthyes)

SHOICHI SUGA[1] and YASUHIKO TAKI[2]

Summary. Previous studies by the authors have revealed that fluoride and iron concentrations in the enameloid of the teeth of some teleostean fishes are very high compared with those in human enamel, and that their concentrations are related to the phylogeny of the fishes rather than to the environmental water, tooth morphology, and feeding habits. In order to trace the origin of the concentration-mechanism of these elements into the enameloid in the evolutionary history of the bony fishes, electron microprobe investigations of tooth fluoride and iron were made in this study in non-teleostean bony fishes which are ancestral to or distantly diverged from the teleostean lineage, i.e., the dipnoan lungfishes, the crossopterygian coelacanth, the primitive antinopterygian polypterids (bichers and reedfishes), gars, and bowfin.

A high fluoride concentration in the entire layer of the enameloid, as seen in many teleostean fishes, was found only in the bowfin, in which the concentration was about 3.5% in the surface layer and about 2.2% in the inner layer. Fluoride concentration was from 0.17% to 0.70% in the enamel of lungfishes and coelachanth and in the enameloid of polypterids and gars. On the other hand, a high iron deposition at the outer layer of enameloid (0.5%–1.3%) was detected in polypterids, gars, and bowfin, while the enamel in lungfishes and coelacanth did not contain any detectable amounts of iron under the analytical conditions of this study.

Based on these results, the mechanisms of high fluoride and iron concentrations seem to have been acquired only in actinopterygians, and the origin of the high fluoride concentration could be traced back to the pre-teleostean stage of actinopterygian evolution while the high iron concentration mechanism seemed to originate as far back as the cladistian (polypterid) stage.

Key words: Teeth—Enameloid—Primitive bony fishes—Fluoride—Iron

[1] Department of Pathology, The Nippon Dental University School of Dentistry at Tokyo, Tokyo, 102 Japan

[2] Tokyo University of Fisheries, Tokyo, 108 Japan

Introduction

Previous investigations on the fluoride (F) concentration in fish teeth using the electron microprobe analysis first indicated that the teleostean fishes may be classified into 2 groups in terms of the fluoride concentration in the enameloid: 1 group whose enameloid contains more than about 2.50% fluoride and the other which contains less than about 0.17%. Such a remarkable difference in fluoride concentration in the enameloid has been found to be related to the phylogeny of the fishes rather than to their feeding habits and the fluoride concentration in the environmental water [1–6], although marine water everywhere in the world contains about 1.3 ppm fluoride [7], while fresh water has usually less than about 0.3 ppm.

Iron (Fe) is also deposited at a high level in the enameloid of some teleostean fishes [8–14]. It was found by the quantitative electron microprobe analysis that the iron concentration in the enameloid of some teleostean fishes ranges from 0.4% to 13.5% [13, 15], whereas that in human enamel from 0.02% to 0.028% in surface enamel and from 0.012% to 0.022% in subsurface enamel [16]. It was suggested that iron concentration is also related to the phylogeny of teleostean fishes [15], although it has been proposed that it is related to the feeding habits of fishes [14].

A comparative investigation using the electron microprobe made on the developing tooth germs of a perciform fish, whose enameloid contains high fluoride and iron concentrations, indicated that the timing and mode of deposition of fluoride and iron into the developing enameloid are different [13].

In order to verify the relationship between the fluoride and iron concentrations in the enameloid and fish evolution more precisely, a quantitative analysis of the fluoride and iron of the teeth of primitive bony fishes was conducted in this study. The classification and interrelationships of major groups of bony fishes (class Osteichthyes) have long been a subject of debate [17–22]. In this study, we tentatively classified the osteichthyans into three subclasses, i.e., the Dipnoi (lungfishes), the Crossopterygii (lobefins, including coelacanths), and the Actinopterygii (ray-finned fish), with the Actinopterygii being further divided into the Cladistia (the Polypteridae or bichers and reedfish, regarded by some authors as representing a distinct subclass, the Brachiopterygii), Chondrostei (sturgeons) and Neopterygii [19–22]. The Neopterygii contains the Ginglymodi (gars) and the Halecostomi, consisting of the Halecomorphi (bowfin) and Teleostei (a major group of advanced actinopterygians, typical familiar fishes ranging from herrings to perches, and the subject of our previous investigations).

Materials and Methods

The teeth examined in the present study were obtained from 3 species of dipnoan lungfishes (Ceratodontidae, Lepidosirenidae, and Protopteridae), 1 species of the crossopterygian coelacanths (Latimeriidae), 5 species of the

Table 1. Fluoride and iron concentrations in the teeth of major groups of bony fishes (class Osteichthyes) other than the Teleostei. Taxa lower than subclass are not given taxonomic categories

Classification/species	*Fluoride (%)*		*Iron (%)*	
Subclass Dipnoi	*(Enamel)*	*(Dentin)*	*(Enamel)*	*(Dentin)*
Neoceratodus forsteri (Australian lungfish)	0.18 ± 0.01	0.20 ± 0.02 p 0.20 ± 0.02	ND	Tr p ND
Lepidosiren paradoxa (South American lungfish)	0.21 ± 0.02	0.19 ± 0.03 p 0.32 ± 0.02	ND	ND p ND
Protopterus dolloi (African lungfish)	0.17 ± 0.01	0.24 ± 0.01 p 0.19 ± 0.02	ND	ND p ND
Subclass Crossopterygii	*(Enamel)*	*(Dentin)*	*(Enamel)*	*(Dentin)*
Latimeria Chalumnae (Coelacanth)	0.34 ± 0.04	0.37 ± 0.01	ND	ND
Subclass Actinopterygii *Cladistia*	*(Enameloid)*	*(Dentin)*	*(Enameloid)*	*(Dentin)*
Polypterus ornatipinnis (Bicher)	a 0.22 ± 0.02 b 0.18 ± 0.01	0.19 ± 0.02	a 0.5 ± 0.3 b ND	ND
Polypterus delnezi (Bicher)	a 0.18 ± 0.02 b 0.19 ± 0.03	0.25 ± 0.02		ND
Polypterus senegalus (Bicher)	a 0.20 ± 0.01 b 0.18 ± 0.02	0.20 ± 0.01	a 2.1 ± 0.9 b ND	ND
Polypterus palmas (Bicher)	a 0.24 ± 0.04 b 0.20 ± 0.02	0.22 ± 0.01	a 1.2 ± 0.2 b ND	ND
Erpetoichthys calabaricus (=*Calamoichthys*, reedfish)	a 0.27 ± 0.08 b 0.21 ± 0.02	0.27 ± 0.02	a 1.0 ± 0.2 b ND	ND
Chondrostei Sturgeons	*Teeth rudimentary or absent*			
Neopterygii Ginglymodi *Lepisosteus spatula* (Gar)	a 0.80 ± 0.01 b 0.20 ± 0.02	0.24 ± 0.01	a 0.7 ± 0.2 b ND	ND
Lepisosteus osseus (Gar) Halecostomi Halecomorphi	a 0.74 ± 0.18 b 0.35 ± 0.12	0.25 ± 0.02	a 0.7 ± 0.3 b ND	ND ND
Amia calva (Bowfin)	a 3.38 ± 0.06 b 2.21 ± 0.03	0.34 ± 0.02	a 0.6 ± 0.2 b ND	ND

a, The outer layer; *b*, the inner layer; *ND*, not detectable; *p*, Pleromin

cladistian bichirs and reedfish (Polypteridae), 2 species of the neopterygian gars (Lepisosteidae), and 1 species of neopterygian bowfin (Amiidae) (Table 1). All these species are freshwater in distribution except for the coelacanth, *Latimeria chalumnae*. They were stored in isopropyl alcohol and/or 10% formalin. Adult sturgeons (Chondrostei) do not have teeth, although young ones have very small teeth which could not be used for the present analysis.

All the specimens were embedded in polyester resin after dehydration in an ascending concentration series of alcohol. Plano-parallel ground sections about 50–60 μm in thickness were made using a grindstone after sectioning with a diamond saw. In order to visualize the mineralization pattern, the ground sections were first microradiographed using soft X-ray (Softex type-CMR, Softex Co., Tokyo). Then, they were mounted on a polished surface of an aluminum block with epoxy resin. The electron microprobe (quantitative and line-scan analyses were performed on the polished surface of the ground sections after applying a thin coating of carbon to form a conducting layer by evaporation in a vacuum.

The electron microprobe apparatus used for the present investigation was a Shimadzu-ARL, type EMX-SM, equipped with a wave length dispersive monochrometer. Simultaneous line-scan analysis of fluoride, calcium, and magnesium with an accelerating voltage of 15 kV and a specimen current of 0.15 μA was carried out on the line which ran perpendicularly towards the enameloid or enamel surface from the dentinal side. Following that, another line-scan analysis for iron, calcium, and magnesium with 20 kV and 0.05 μA was performed along the same line, in order to compare the patterns of fluoride and iron distributions. A 10.16 cm RAP curved single crystal was used for the detection of the FKα and CaKβ emissions, a 10.16 cm LiF for the FeKα emission, and a 10.16 cm ADP for CaKα and MgKα emissions.

Quantitative analysis was performed by measuring the intensity of FKα and FeKα emissions obtained by the mode of point analysis using fluorapatite (Durango, F: 3.5%) and the glass beads of anhydrous lithium tetraborate ($Li_2B_4O_7$) containing various amount of iron (III) oxide (Fe_2O_3) (Fe: 5.0%, 3.0%, 1.0%, 0.5%, 0.1%, and 0.05%), which were fused in a furnace (1100°C)

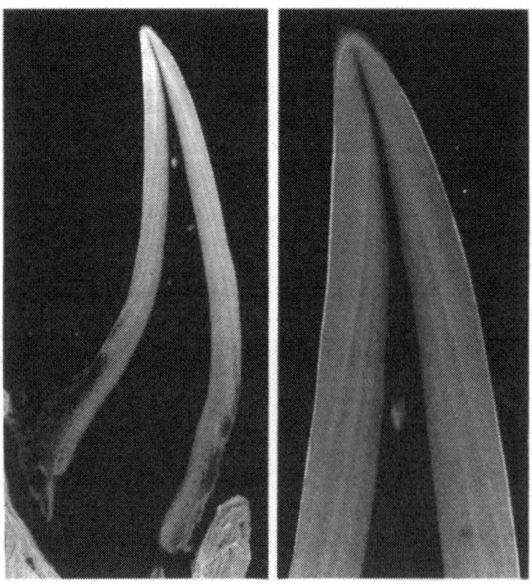

a b

Fig. 1. Microradiograms of a longitudinal ground section of the tooth of a coelacanth, *Latimeria chalumnae*. **a** Low magnification showing the entire view of a tooth. **b** High magnification of the upper one-third of **a**. The tooth surface is covered by very thin highly mineralized enamel

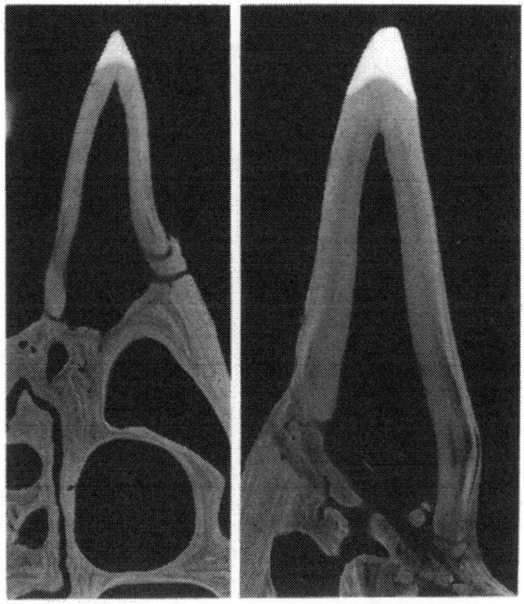

a b

Fig. 2. Microradiograms of longitudinal ground sections of the teeth of cladistian bicher and reedfish. **a** *Polypterus senegalus* (bicher). **b** *Calamoichthys calabaricus* (reedfish)

[23], as standard samples. A point analysis was performed at certain points in the enameloid or enamel and the dentin. The patterns of line-scan analysis made from the dentin layer towards the enameloid or enamel surface were compared with the mineralization pattern revealed by microradiography.

Results

Form, Structure and Mineralization Pattern of Teeth

Coelacanth, polypterids, gars, and bowfin have separate chisel-like teeth. In the microradiogram of longitudinal ground sections of the teeth of these fishes, the hard tissue, which can be very easily distinguished from the dentin by its very high degree of mineralization, was observed to cover the coronal surface of teeth (Figs. 1–3). In the teeth of polypterids, gars and bowfin, the arrowhead-shaped highly mineralized tissue, which has been considered to be the enameloid, covered only the cuspal tip of the teeth (Figs. 2, 3). Alternatively, in the teeth of the coelacanth, a very thin highly mineralized tissue, which has been identified as the enamel [24–26], covered the entire coronal surface of the teeth (Fig. 1).

The 3 species of lungfishes have plate-like teeth. In the microradiograms of longitudinal ground sections, a wide central part of dentin adjacent to the pulpal space was more highly mineralized than the rest of the dentin and bone. It has been suggested that this highly mineralized part be named "pleromin" or "petrodentin" because of the peculiar structure of the organic matrix and the mineralization pattern [27]. On the other hand, the lateral thin layer of

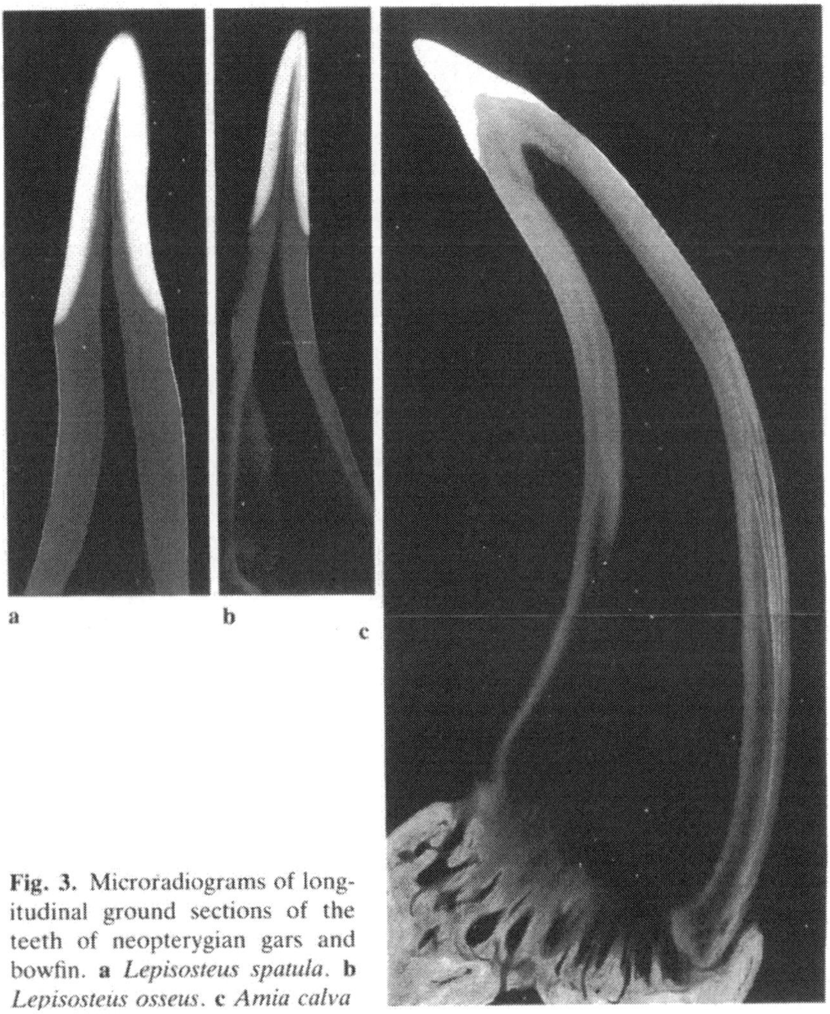

Fig. 3. Microradiograms of longitudinal ground sections of the teeth of neopterygian gars and bowfin. **a** *Lepisosteus spatula*. **b** *Lepisosteus osseus*. **c** *Amia calva*

the tooth plate is also very highly mineralized (Fig. 4). This layer has been suggested to be enamel by some investigators [28–30], although others considered it to be enameloid [31, 32]. In the present investigations, this highly mineralized layer is tentatively called the "enamel."

Elemental Analysis by the Electron Microprobe

In all the fishes examined, the line-scan analysis performed along a line crossing the enameloid or enamel and dentin layers of the fully mineralized teeth showed that the calcium concentration is higher in the enameloid or enamel than in the dentin (Figs. 5–9).

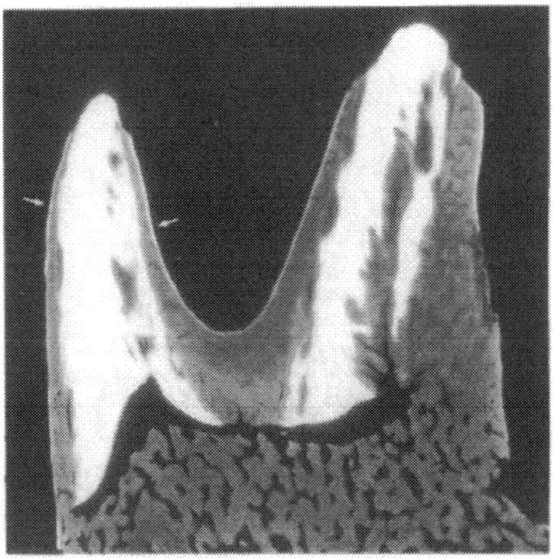

Fig. 4. Microradiogram of a longitudinal ground section of the tooth plate of a dipnoan lungfish, *Lepidosiren paradoxa* (South American lungfish). *Arrow* indicates the highly mineralized thin enamel

Magnesium concentration is much lower in the enameloid and enamel than in the dentin, and a sharp decline of magnesium concentration was observed at the enameloid (or enamel)-dentin junction, followed by a gradual decrease towards the enameloid or enamel surface (Figs. 5–9). In the pleromin of the lungfishes, the calcium concentration was higher and the magnesium concentration much lower than in the rest of the dentin (Fig. 5).

Fluoride concentration was very low and showed an almost even distribution throughout the entire layer of enamel or enameloid and dentin of the teeth of the lungfishes, coelacanth, and polypterids (Figs. 5a, 7a); it was also very low in the pleromin of lungfishes (Fig. 5a). In the teeth of the gars, the fluoride concentration was very low in the dentin and enameloid, except for the narrow outer layer in which fluoride showed a slight increase of its concentration towards the surface (Fig. 8a). In the teeth of the bowfin, fluoride concentration was also very low in the dentin; however, as it entered into the enameloid, the fluoride at first increased its concentration very sharply and highly and then increased gradually towards the surface (Fig. 9a).

A quantitative point analysis performed at the outer and inner layers of the enameloid or enamel and at the middle layer of dentin indicated that the fluoride concentration ranged from about 0.17% to 0.34% in the entire layer of the enameloid or enamel, from about 0.19% to 0.37% in the dentin of the lungfishes, coelacanth, and polypterids, and from about 0.19% to 0.32% in the pleromin of the lungfishes. However, in the enameloid of the gars, the fluoride concentration was about 0.80% and 0.74% in the outer layer and about 0.20% and 0.35% in the inner layer, and, in the enameloid of the bowfin, it was about 3.38% in the outer layer and about 2.21% in the inner layer. In the dentin of the gars and bowfin, the fluoride concentration ranged from about 0.24% to 0.34% (Table 1).

Line-scan analysis indicated that the iron concentration was very low throughout the entire layer of enamel and dentin of the lungfishes (Fig. 5b) and

a b

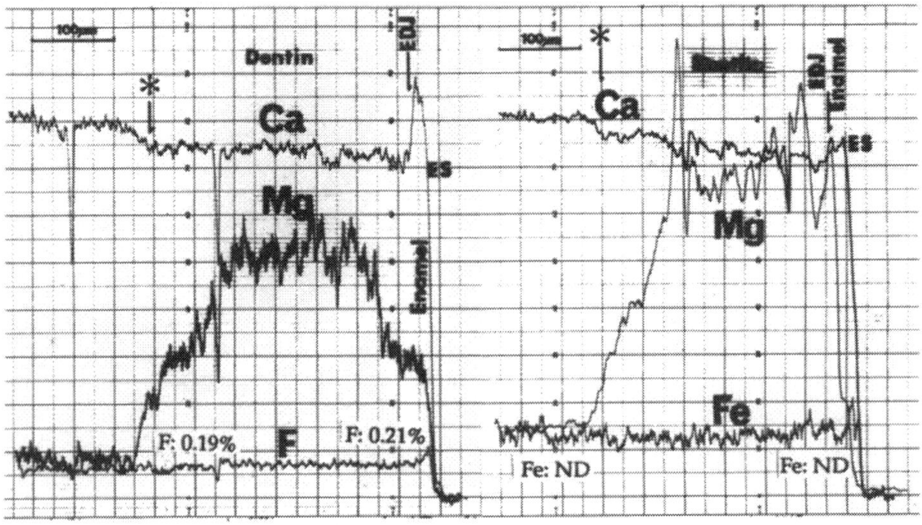

a b

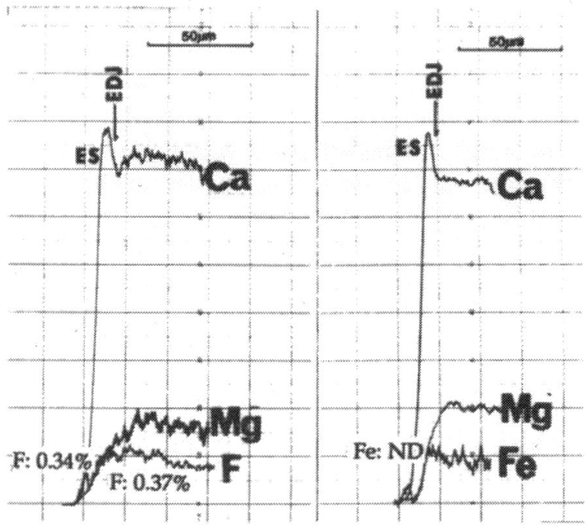

Fig. 5. (*Top*) The distributions of fluoride, calcium, magnesium, and iron in the tooth of a lung fish, *Lepidosiren paradoxa*, revealed by line scan analyses by means of the electron microprobe, which were performed along the same line crossing the enamel, dentin and pleromin. The results of a quantitative point analysis of fluoride and iron at the certain points are indicated. **a** The distribution of fluoride, calcium and magnesium. **b** The distribution of iron, calcium and magnesium. *EDJ*, Enamel-dentin junction; *ES*, enamel surface, *, dentin-prelomin junction. Pleromin is to the *left* of the *

Fig. 6. (*Bottom*) The distribution of fluoride, calcium, magnesium, and iron in the tooth of a coelacanth, *Latimeria chalumnae*, revealed by line-scan analyses by means of the electron microprobe, which were performed along the same line crossing the enamel and dentin. Results of a quantitative point analysis of fluoride and iron at the certain points are indicated. **a** Distribution of fluoride, calcium, and magnesium. **b** Distribution of iron, calcium, and magnesium. *EDJ*, Enamel-dentin junction; *ES*, enamel surface. The enamel is between ES and EDJ and the dentin is *to the right* of EDJ the coelacanth (Fig. 6b), and in the pleromin of the lungfishes (Fig. 5b). According to a quantitative point analysis, the emission intensity of iron could not be discriminated from the background value (Table 1). However, in the enameloid, especially in its outer layer, of polypterids, gars, and bowfin, iron concentration showed a steep increase from the side of the middle layer towards the surface, whereas, in the rest of enameloid and dentin, it was very low (Figs. 7b–9b). A quantitative point analysis indicated that iron concentration ranged from 0.5% to 1.3% at the outer layer and from about 0.1% to 0.2% at the middle layer of the enameloid of these fishes, whereas, in the inner layer of enameloid and in the dentin, it was not detectable under the present analytical condition (Table 1).

Discussion

A quantitative electron microprobe fluoride analysis of the teeth of the Perciformes, which is one of the largest orders of teleostean fishes and includes a considerable number of freshwater species in addition to the predominant marine forms, showed that, while the dentin and bone contained fluoride lower than about 0.50%, the enameloid of all of the 78 species examined contained in its surface layer a rather high fluoride level, ranging mainly from 2.10% to 5.45% [6]. There was no significant difference in the fluoride concentration associated with the saline environments of the water in which these fishes live.

The teeth of marine species of the Tetraodontiformes, which are considered to have evolved from a perciform-like stem and are classified into 2 suborders, i.e., the more generalized Balistoidei and the more specialized Tetraodontoidei, were also examined in the same way [4]. The results indicated that the enameloid of the balistoids studied contained a very high fluoride concentration, ranging from 2.31% to 5.16%, whereas the concentration in the tetraodontids was only 0.14%–0.28%. In the perciform family Acanthuridae (marine fishes) from which the Teraodontiformes may have stemmed, the fluoride concentrations were higher than 3.0% [4]. These facts suggest that the fluoride concentration in the enameloid of teleostean fishes is related to the phylogeny of the fishes rather than to the fluoride concentration in the environmental water, and that the fluoride concentration mechanisms may have been lost secondarily in some groups. However, the biological significance of the high concentration of fluoride into the enameloid is not clearly understood.

Crystallographic investigations indicated that fluoride concentration into the enameloid results in a greater crystal size, reduction in carbonate content, and systematic decrease in the *a*-axis lattice parameter [33].

Balistoid, tetraodontoid, and acanthurid fishes have also been subjected to iron analysis using the electron microprobe [15]. The results showed that the balistoids and acanthurids contain a rather high iron concentration in the enameloid, in the former at the surface layer (0.4%–13.5%) and in the latter throughout the entire layer (2.7%–3.9%), but the tetraodontoids contain such a

b

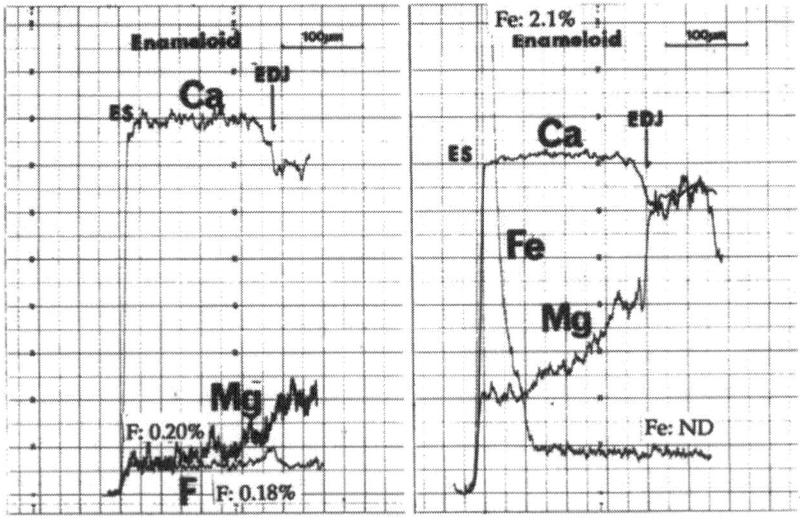

Enameloid
100µm
EDJ
Ca
ES
Mg
F: 0.20%
F
F: 0.18%
Fe: 2.1%
Enameloid
100µm
Ca
EDJ
ES
Fe
Mg
Fe: ND

a

b

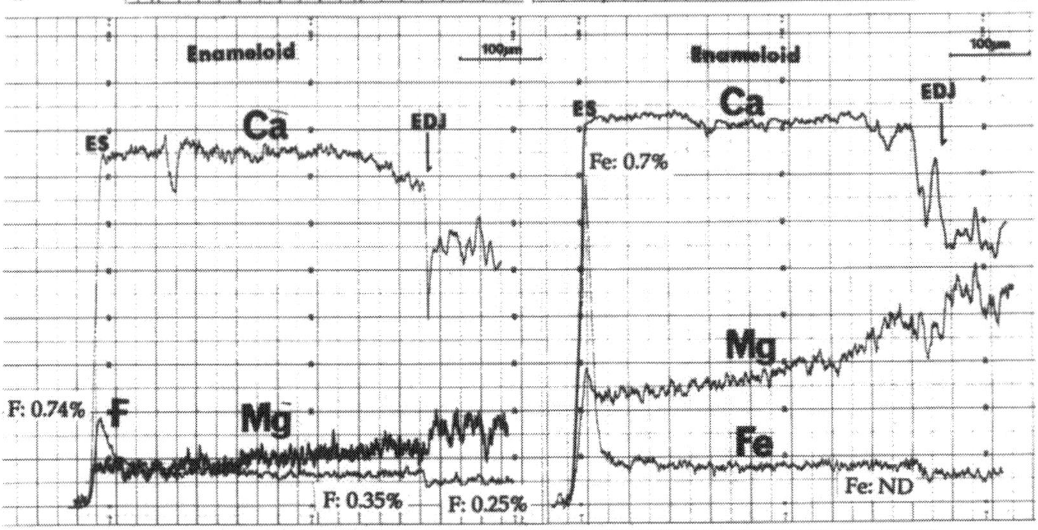

Enameloid
100µm
Ca
EDJ
ES
F: 0.74%
F
Mg
F: 0.35%
F: 0.25%
Enameloid
100µm
Ca
EDJ
ES
Fe: 0.7%
Mg
Fe
Fe: ND

a

b

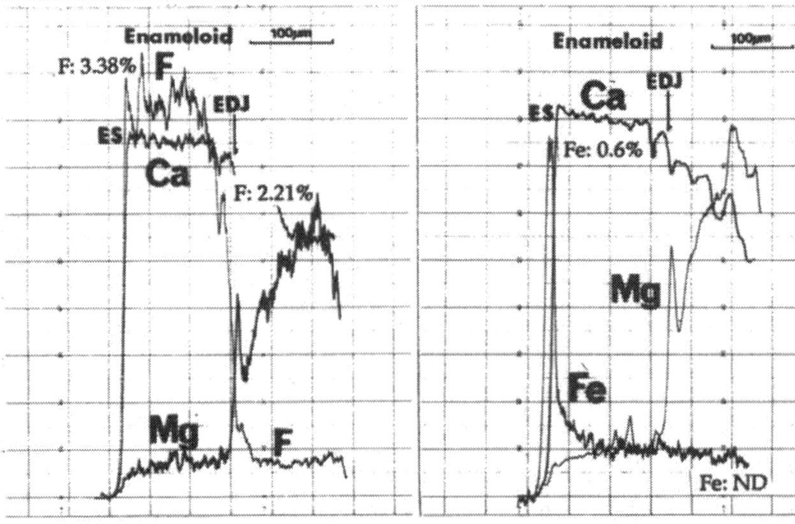

Enameloid
100µm
F: 3.38%
F
EDJ
ES
Ca
F: 2.21%
Mg
F
Enameloid
100µm
Ca
EDJ
ES
Fe: 0.6%
Mg
Fe
Fe: ND

a very low iron concentration that it could not be discriminated from the background value. It is inferred from these results that the iron concentration into the enameloid is also related to the phylogeny of these fishes and that the mechanism has been lost in the course of differentiation of the tetraodontoids. However, some investigators consider that the iron deposition hardens the enameloid and reduces abrasion and cracking, and is related to the feeding habits of fishes [12, 14].

The chemical state and significance of the iron deposition in the enameloid have not been elucidated. However, the presence of $Fe3+$ ions is known to induce a high crystallinity of apatite [34]. Electron microscopic investigations of heavily deposited iron at the narrow surface layer of enamel of the rat incisor showed that it is not present as a separate particulate phase but is mainly bound to the surface of apatite crystals [35]. A Mössbauer effect study also suggested that iron is mainly attached on the surface of crystals [36].

It is interesting to note that both fluoride and iron, which originally have no particular chemical correlation, are deposited together in some heavily mineralized tissues, such as the enameloid of some teleostean fishes [4, 13, 15], and the teeth of the radula of the chiton (Polyplacophora: Mollusca) [13, 37], although their depositions into the enameloid are not fully synchronized with each other in the process of mineralization of the developing enameloid and their final distributional patterns do not coincide [13]. At least in the enameloid of teraodontiform fishes such a phenomenon seems to be related to the phylogeny of the fishes [15], in view of their concurrent disappearance in the course of differentiation of these fishes.

←

Fig. 7. (*Top*) The distribution of fluoride, calcium, magnesium, and iron in the tooth of a bicher, *Polypterus senegalus*, revealed by line-scan analyses by means of the electron microprobe, which were performed along the same line crossing the enameloid and dentin. Results of a quantitative point analysis of fluoride and iron at the certain points are indicated. **a** Distribution of fluoride, calcium and magnesium. **b** Distribution of iron, calcium and magnesium. *EDJ*, Enameloid-dentin junction; *ES*, enameloid surface. The dentin is *to the right* of EDJ

Fig. 8. (*Middle*) The distribution of fluoride, calcium, magnesium, and iron in the tooth of a gar, *Lepisosteus spatula*, revealed by line-scan analyses by means of the electron microprobe performed along the same line crossing the enameloid and dentin. Results of a quantitative point analysis of fluoride and iron are indicated. **a** The distribution of fluoride, calcium, and magnesium. **b** The distribution of iron, calcium, and magnesium. *EDJ*, Enameloid-dentin junction; *ES*, enameloid surface. The dentin is *to the right* of EDJ

Fig. 9. (*Bottom*) The distribution of fluoride, calcium, magnesium, and iron in the tooth of bowfin, *Amia calva*, revealed by line-scan analyses by means of the electron microprobe, which were performed along the same line crossing the enameloid and dentin. Results of a quantitative point analysis of fluoride and iron are indicated. **a** The distribution of fluoride, calcium and magnesium. **b** The distribution of iron, calcium and magnesium. *EDJ*, Enameloid-dentin junction; *ES*, enameloid surface. The dentin is *to the right* of EDJ

All the above observations point to the existence of certain relationships between fluoride and iron concentrations and the phylogeny of teleostean fishes. In order to elucidate these relationships more precisely and trace the origin of the concentration mechanisms, the teeth of non-teleostean bony fishes ancestral to or distantly divergent from the present-day predominating teleostean fishes were examined in this study using the same methods as in our previous studies [1–6, 13, 15].

The coronal portion of the teeth of all bony and cartilaginous fishes had been considered to be covered with enameloid, which is very highly mineralized, as observed in the enamel of tetrapods, and in which very fine filaments of collagen formed by odontoblasts persist; there is also a complex woven structure with varying orientations of the crystals [12, 29, 38]. However, recent histological investigations of fish teeth using transmitted, polarizing, and scanning electron microscopies, and microradiography have revealed that the teeth of the coelacanth are covered with true enamel, in which the crystals are arranged perpendicularly to the surface without any trace of collagen and with incremental lines growing away from the side of the dentin and running throughout the entire layer [24, 25]. This fact was considered to be of great importance since some crossopterygians have been considered by many authors as being ancestral to tetrapods [24]. Properties similar to the enamel, rather than the enameloid, containing crystals that run perpendicularly to the surface, have also been observed in lungfishes [28–30], although other investigators considered them to be enameloid or highly mineralized dentin [31, 32, 39]. On the basis of cladistic analyses of anatomical characters, it was indicated that lungfishes and tetrapods are sister groups [21, 40, 41].

The present study revealed that no high concentrations of fluoride and iron, as observed in the enameloid of some teleosteans, were found in the enamel of the coelacanth and lungfishes, whereas, calcium and magnesium in these 2 groups showed a distributional pattern similar to that in the enamel of mammals and in the enameloid of fishes [4–6, 15, 42].

The relationships of the cladistians, or the polypterids, have been the subject of a century-long dispute. They have been considered to represent either a highly specialized lineage of the subclass Actinopterygii (ray-finned fish) or a group sufficiently distinct from other osteichthyans to warrant their separation into a subclass, the Brachiopterygii [18, 22, 43]. The fact that the polypterids have enameloid is in favor of, or at least not inconsistent with, their allocation to the Actinopterygii.

In the present study, high concentrations of iron in the enameloid, especially in its surface layer, were observable in the polypterids, gars, and bowfin, whereas the adjacent dentin contained a very low iron concentration which was not detectable under the analytical conditions of this study. The concentration of the iron in the enameloid of these fishes increased steeply from the middle layer towards the surface, as seen in many teleosteans [13, 15]. On the other hand, very high concentrations of fluoride throughout the entire layer of enameloid, as seen in many teleosteans, were found only in the bowfin. In the other examined species, only the neopterygian gars showed a slight increase in the narrow outer layer.

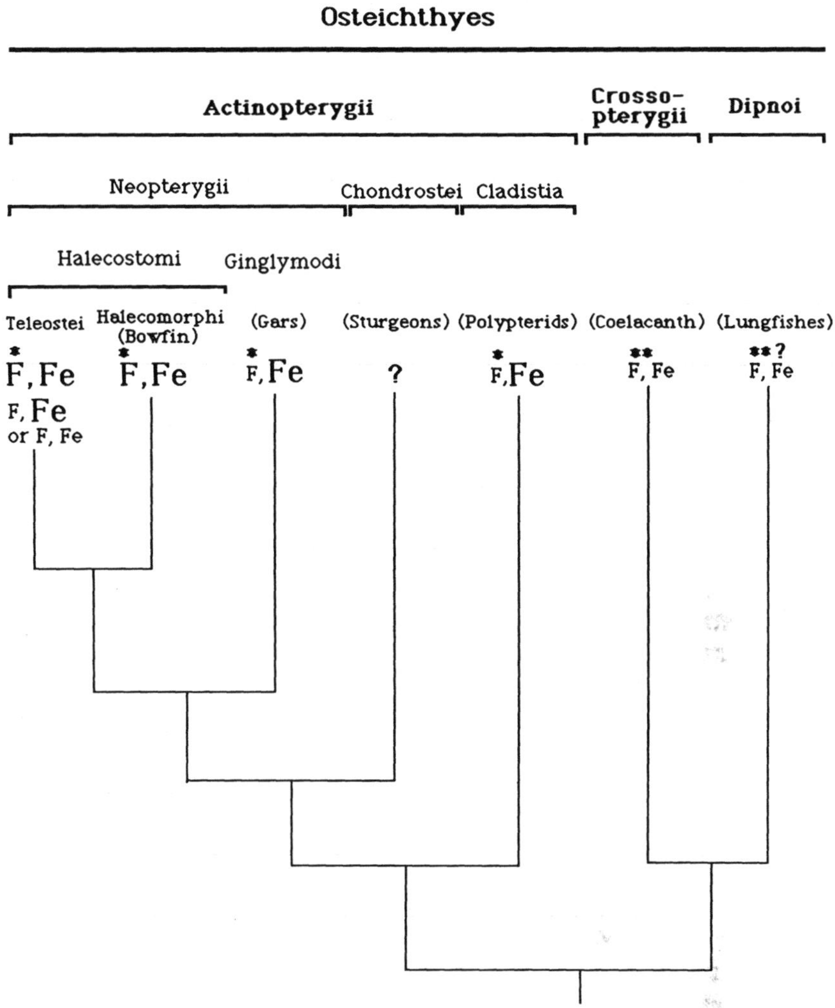

Fig. 10. A possible phylogenetic diagram of major groups of extant bony fishes (Osteichthyes) with summarized results of the quantitative analysis of fluoride and iron in the enameloid and enamel. * Enameloid, ** Enamel, ***F, Fe***, High concentration in the enamel and enameloid (fluoride = >3.0%; iron = >0.5%); F, Fe, low concentration in the enamel and enameloid (fluoride = <0.4%; iron = not detectable under the present analytical conditions)

Incorporating all the observations made in this study, the mechanisms of high fluoride and iron concentrations into the developing enameloid are considered to have occurred only in the Actinopterygian lineage. The high iron concentration in the polypterids supports their inclusion in the Actinopterygii. The mechanism of high iron concentration can be traced back to the cladistian stage of actinoperygian evolution, while that of high fluoride concentration to the neopterygian stage (Fig. 10). The two mechanisms are probably independent of each other and have different evolutionary backgrounds. This idea is supported by the fact that the iron concentration is very low in the enameloid of sharks and rays of cartilaginous fishes, whereas the fluoride concentration is very high [44]. Since, in addition to the present results, it was found that there are some teleostean fishes of the Characiformes whose enameloid contains very high iron and low fluoride concentrations [44], the high iron concentration mechanism seems to have appeared earlier and disappeared later than the high fluoride concentration mechanism in the actinopterygian evolution.

Deposition of histologically or histochemically detectable amounts of iron has been found in the enameloid of some teleostean fishes and in the enamel of some amphibians, reptiles, and mammals, although no investigation has been made on its biological and phylogenetic significance [8–10, 12]. In view of the absence of detectable amounts of iron in the enamel of the Dipnoi and Crossopterygii, the iron deposition mechanisms of tetrapods may have appeared in parallel with that of actinopterygian fishes, in the former group after they established themselves as terrestrial animals and in the latter from the cladistian stage.

High fluoride concentrations are not found in amphibians, reptiles, or mammals [45] nor in dipnoans or crossopterygians. The high concentration of fluoride seen in the enameloid of neopterygian fishes is probably due to a certain mechanism specially designed for advanced actinopterygians for some unknown biological reasons.

Acknowledgements. We thank Dr. T. Ueno, National Science Museum, Tokyo, and Dr. D.E. McAllister, National Museum of Natural Sciences, Ottawa, for supplying many specimens.

References

1. Suga S, Wada K, Ogawa M (1977) Fluoride concentration in the enameloid of fishes (abstract). J Dent Res 56(A):54
2. Suga S, Wada K, Ogawa M (1978) Possible relationship between the fluoride concentration in the enameloid and phylogeny of fishes (abstract). J Dent Res 57(A):280
3. Suga S, Wada K, Ogawa M (1980) Fluoride concentration in the enameloid of fishes. In: Omori M, Watabe N (eds) The mechanisms of biomineralization in animals and plants. Tokai University Press, Tokyo, pp 229–240
4. Suga S, Wada K, Ogawa M (1981) Fluoride concentration in teeth of tetraodontiform fishes and its phylogenetic significance. Jpn J Ichthyol 28:304–312
5. Suga S, Wada K, Ogawa M (1981) Fluoride concentration in the enameloid of fish teeth and its relationship to the phylogeny of fishes. In: Binder K, Hohenegger M (eds) Fluoride metabolism. Whilhelm Maudrich, Wien, pp 79–88

6. Suga S, Taki Y, Wada K (1983) Fluoride concentration in the teeth of perciform fishes and its phylogenetic significance. Jpn J Ichthyol 30:81–93
7. Greenhalgh R, Riley JP (1963) Occurrence of abnormally high fluoride concentration at depth in the ocean. Nature 197:371–372
8. Schmidt WJ (1958) Natürliche Färbung von Reptilien- und Fischzähnen durch Eisenoxyd. Zool Anz 161:168–178
9. Schmidt WJ (1969) Die Schmelznatur der eisenoxidhaltigen Kappe auf Teleostier-Zähnen. Z Zellforsch Mikrosk Anat 93:447–450
10. Miles AEW (1963) Pigmented enamel. Proc R Soc Med 56:918–920
11. Garant PR (1970) Observations on the ultrastructure of the ectodermal component during odontogenesis in *Helostoma temmincki*. Anat Rec 166:167–188
12. Shellis RP, Berkovitz BKB (1976) Observations on the dental anatomy of piranhas (Characidae) with special reference to tooth structure. J Zool 180:69–84
13. Suga S (1984) The role of fluoride and iron in mineralization of fish enameloid. In: Fearnhead RW, Suga S (eds) Tooth enamel IV. Elsevier, Amsterdam, pp 472–477
14. Motta PJ (1987) A quantitative analysis of ferric iron in butterflyfish teeth (Chaetodontidae, Perciformes) and the relationship to feeding ecology. Can J Zool 65:106–112
15. Suga S, Wada K, Taki Y, Ogawa M (1989) Iron concentration in teeth of tetraodontiform fishes and its phylogenetic significance. J Dent Res 68: 1115–1123
16. Torell P (1957) Determination of iron in dental enamel. Odont Tidskr 65:20–23
17. Nelson GJ (1969) Gill arches and the phylogeny of fishes, with notes on the classification of vertebrates. Bull Am Mus Nat Hist 141:475–552
18. Jessen HL (1973) Interrelationships of actinopterygians and brachiopterygians: Evidence from pectral anatomy, In: Greenwood PH, Miles RS, Patterson C (eds) Interrelationships of fishes. Academic, London, pp 227–232
19. Patterson C (1973) Interrelationships of holosteans. In: Greenwood PH, Miles RS, Patterson C (eds) Interrelationships of fishes. Academic, London, pp 233–305
20. Shaeffer B (1973) Interrelationships of chondrosteans. In: Greenwood PH, Miles RS, Patterson C (eds) Interrelationships of fishes. Academic, London, pp 207–226
21. Rosen DE, Forey RL, Gardiner BG, Patterson C (1981) Lungfishes, tetrapods, paleontology, and plesiomorphy. Bull Am Mus Nat Hist. 167: 159–276
22. Lauder GV, Liem KF (1983) The evolution and interrelationships of the actinopterygian fishes. Bull Mus Comp Zool 150:95–197
23. Stephenson DA (1969) An improved flux-fusion technique for X-ray emission analysis. Anal Chem 41:966–967
24. Smith MM (1978) Enamel in the oral teeth of *Latimeria chalumnae* (Pisces: Actinistia): A scanning electron microscope study. J Zool Lond 185:355–369
25. Shellis RP, Poole DFG (1978) The structure of the dental hard tissues of the coelacanthid fish *Latimeria Chalumnae* Smith. Arch Oral Biol 23:1105–1113
26. Sasagawa I, Ishiyama M, Kodera H (1984) Fine structure of the pharyngeal teeth in the coelacanthid fish (*Latimeria chalumnae*). In: Fearnhead RW and Suga S (eds) Tooth enamel IV. Elsevier, Amsterdam, pp 462–466
27. Orvig T (1976) Palaeohistological notes. 4. The interpretation of osteodentine, with remarks on the dentition in the devonian dipnoan griphognathus. Zool Scripta 5:79–96
28. Peyer B (1968) Comparative odontology, University of Chicago Press, Chicago, pp 110–118
29. Schmidt WJ, Keil A (1971) Polarizing microscopy of the dental tissues. Pergamon, Oxford, pp 240–244

30. Ishiyama M, Ogawa T (1983) Existence of true enamel on tooth plate in the lungfish *Lepidosiren paradoxa* (in Japanese). Acta Anat Nipponica 58:157–161
31. Smith MM (1977) The microstructure of the dentition and dermal ornament of three dipnoans from the Devonian of Western Australia. Phil Trans R Soc (B) 281:29–72
32. Kemp A (1979) The histology of tooth formation in the Australian lungfish, *Neoceratodus forsteri Krefft*. Zool J Linn Soc 99:251–287
33. LeGeros RZ, Suga S (1980) Crystallographic nature of fluoride in enameloids of fish. Calcif Tissue Int 32:169–174
34. LeGeros RZ, Tsheri MH, Quirolgico CB, LeGeros JP (1980) Formation and stability of apatite: Effects of some cationic substituents. Proc 2nd International Congress on Phosphorus Compounds, Boston, pp 89–103
35. Selvig AK, Halse A (1975) The ultrastructural localization of iron in rat incisor enamel. Scand J Dent Res 83:88–95
36. Bauminger E, Ofer S, Gedalia I, Horowitz G, Mayer I (1985) Iron uptake by teeth and bones: A Mössbauer effect study. Calcif Tissue Int 37:386–389
37. Lowenstam HA (1967) Lepidocrocite, an apatite mineral, and magnetite in teeth of chitons (Polyplacophora). Science 156:1373–1374
38. Shellis RP (1981) Dental tissues evolution In: Osborn JW (ed) Dental anatomy and embryology. Blackwell Scientific, Oxford, pp 155–165
39. Denison RH (1974) The structure and evolution of teeth in lungfishes. Fieldiana Geology 33:31–58
40. Gardiner BG (1980) Tetrapod ancestry: A reappraisal. In: Panchen AL (ed) The terrestrial environment and the origin of land vertebrates. Academic, London, pp 175–185
41. Forey PL (1986) Relationships of lungfishes. In: Bemis WE, Burggren WW, Kemp NE (eds) The biology and evolution of lungfishes. Alan R Liss, New York, pp 75–91
42. Suga S, Kondo M, Onodera A, Kubota Y, Ohtuka M (1971) Electron microprobe analysis on the distributions of Cl, Mg and Na, in the enamels of various animals (in Japanese). Jpn J Oral Biol 13:85–94
43. Patterson C (1982) Morphology and interrelationships of primitive actinopterygian fishes. Am Zool 22:241–259
44. Suga S, Taki Y, Wada K, Ogawa M (1991) Evolution of fluoride and iron concentrations in the enameloid of fish teeth, In: Suga S, Nakahara H (eds) Mechanisms and phylogeny of mineralization in biological systems. Springer, Tokyo, pp 439–446
45. Suga S, Ogawa M (1989) Comparative observations on trace element distribution in enamel and enameloid, In: Fearnhead RW (ed) Tooth enamel V. Florence, Yokohama, pp 353–359

Fish Tooth Formation: An Assessment of Biological Factors Affecting the Fluoride Content of Enameloid

KENNETH S. PROSTAK, PHILIP SEIFERT, and ZIEDONIS SKOBE[1]

Summary. The fluoride content in fish tooth enameloid is dependent on species phylogeny, not on its food or habitat. The source of osteichthian enameloid collagen, formed from both the ectoderm and mesoderm, has no relationship to the fluoride content of the enameloid. Similarities in the electrophoretic profiles, amino acid composition and the extent of collagen cross-linking between bony fish enameloids having a high or low fluoride content make it unlikely that specific proteins of the enameloid matrix avidly bind fluoride. Chondrichthian enameloid is distinct from that of bony fishes: it contains no ectoderm collagen and little mesodermal collagen, but is enriched in carbohydrate and non-collagenous mesodermal proteins. On the other hand, the unique chondrichthian enameloid suggests a fluoride-binding property.

The serum fluoride levels in all species tested were low, apparently not related to the fluoride content of their enameloid. Therefore, it seems unlikely that there is a systemic mechanism concentrating fluoride into enameloid.

We propose that a triad of cellular morphologies, found exclusively in bony fish tooth buds forming a high fluoride enameloid, provides the structural framework for fluoride concentration into the enameloid. First, a rich vascular network surrounding the enameloid organ in the mineralization stage allows site-specific enrichment of ions, including fluoride, from fenestrated capillaries into enameloid organ cells. Second, the specialized morphology of outer dental epithelial cells in species with high fluoride levels consistent with the hypothesis that fluoride ions are preferentially transported across cell membranes as are chloride ions in "chloride cells" of fish gills. Third, the intercellular junctions of the enameloid organ serve as barriers to the outward diffusion of transported fluoride.

Key words: Enameloid—Development—Fluoride—Ion Transport—Matrix

Introduction

The homology of fish tooth enameloid with mammalian enamel has been debated considerably over the past century. The presence of collagen in the enameloid matrix has led to the perception that this tissue was a modified

[1] Forsyth Dental Center, Boston, MA 02115, USA

dentin [1]. Subsequently, it has been suggested that an epithelial protein secreted into the mesodermal collagen modified the collagen cross-linking, allowing large apatite crystals to form [2]. However, we showed that collagen of the enameloid matrix was not mesodermal in origin, but that the inner dental epithelial cells (IDE) produced a large quantity of type I collagen in the enameloid matrix [3, 4]. IDE cells of several bony fish also secrete an amorphous material on the surface of mineralizing enameloid. Here, we provide additional evidence that the enameloid matrix of the bony fish enameloid is composed primarily of ectodermally derived collagen.

It is known that bony fishes may be classified into 2 groups: the fishes whose enameloid contains high fluoride (higher than about 2.0%) and those whose enameloid is low in fluoride (lower than about 0.4%). The fluoride content in the enameloid is dependent on species phylogeny, not fluoride concentration in the environmental water or the feeding habits of the fishes [5–7; Chap. 2] We demonstrated that the fluoride content of the surface layer reflects the high or low fluoride content of its underlying enameloid. Thus, it is unlikely that these amorphous proteins have specific fluoride-concentrating properties. The observation of mineralizing amorphous material adjacent to the pulpal aspect of enameloid suggests that odontoblasts associated with the enameloid matrix synthesis also synthesize and secrete non-collagenous proteins into the enameloid matrix. The homology of these and other surface layers to mammalian dental enamel is explored.

Little is known concerning the biochemistry of collagen and other proteins of the enameloid matrix, especially in relation to the fluoride content of enameloid. In this study, we determined whether potential differences in electrophoretic gel patterns, amino acid composition, and collagen cross-linking between high and low fluoride-containing enameloid matrices account for the different fluoride content in enameloid.

Finally, data are presented supporting a novel hypothesis of fluoride regulation through the enameloid organ of bony fish containing high fluoride in enameloid. Unique cytostructural features of the outer dental epithelial cells (ODE) are compared to the anion regulating "chloride cells" of fish gills. We suggest that fluoride is transported by specialized ODE cells found in tooth buds forming high fluoride enameloid.

Materials and Methods

Numerous specimens representing the following families were routinely processed for either TEM [4, 8] or freeze-dried for elemental analysis by electron microprobe: Skate-*Rajidae*; Blue Shark-*Galeidae*; Tautog-*Labridae*; Sea Robin-*Triglidae*; Parrotfish-*Scaridae*; Cichlid-*Cichlidae*; Triggerfish-*Balistidae*; Pufferfish-*Tetraodontidae*; Alligator Gar-*Lepisosteidae*; Bluefish-*Pomatomidae*; and Goldfish-*Cyprididae*.

Some fish were injected with either colchicine [4, 8], insulin, or glucagon (2 mg/kg), killed at various intervals post-injection, and then processed for TEM.

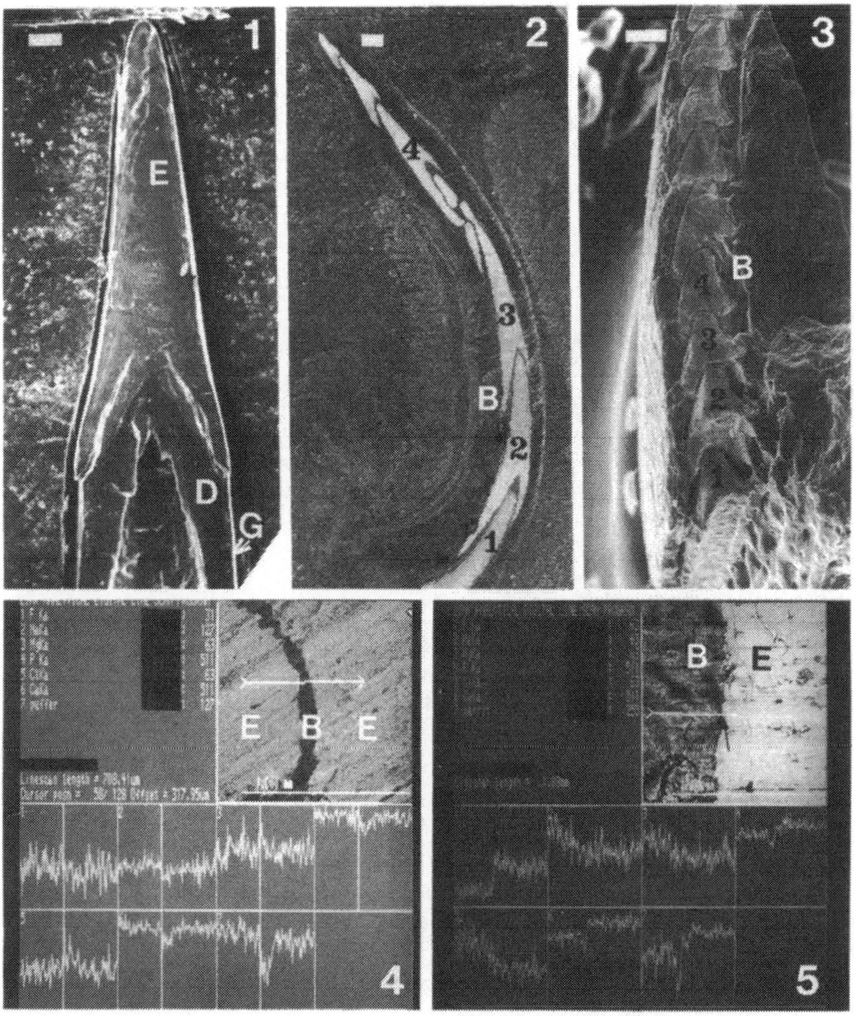

Fig. 1. Secondary electron image of longitudinally sectioned, mature gar tooth depicting enameloid (*E*), dentin (*D*), and ganoid/enamel (*G*). *Scale* = 100 µm

Fig. 2. Backscattered electron image of a parrotfish jaw section showing arrangement of spade-shaped teeth. Bone (*B*), but not dentin, is interspersed between the enameloid pieces (*1–4*) of successive tooth rows. *Scale* = 100 µm

Fig. 3. Secondary electron image of a fractured sharpnose pufferfish jaw showing the eruptive pattern of the plate-like enameloid pieces (*1–4*). Dentin is not present in this species. *B* = Bone. *Scale* = 100 µm

Fig. 4. EDS linescan of mature pufferfish enameloid. The *line* on the digitized image indicates the path of the linescan. The numerical values for each element listed correspond to the total vertical scale CPM. There were distinct decreases in calcium (*6*) and phosphorus (*4*), but not fluoride (*1*), at the intersection of enameloid (*E*) and bone (*B*). *Block 7* is a grey-scale histogram of image. The *arrows* indicate the position of bone

Fig. 5. EDS linescan of mature parrotfish enameloid. Distinct increases in calcium (*6*), phosphorus (*4*), and fluoride (*1*) were observed in enameloid (*E*) relative to bone (*B*). Chloride (*5*) content of bone is higher relative to that in enameloid. *Block 7* is a grey-scale histogram of the image. The *arrows* indicate the position of bone

Environmental water, the developing enameloid matrix, and blood samples of each fish were taken and assayed for fluoride by microdiffusion [9]. The amino acid composition and collagen cross-link content of freeze-dried parrotfish, pufferfish, and blue shark enameloids were determined by Y. Kuboki [10]. Ten-percent polyacrylamide gel electrophoresis was performed on 0.5 M acetic acid soluble enameloid proteins of several developing tooth stages from tautog, parrotfish, pufferfish, skate, and shark. Antisera developed against the rat enamel matrix [3] was used in Western blotting and immunocytochemical [11] assays to determine which enameloid proteins were potentially homologous to those of the mammalian enamel matrix.

Morphometric analyses were performed on calibrated micrographs using either SigmaScan (Jandel Scientific; Calif.) or Link (Link Analytical; Wis.) analytical software.

Results and Discussion

Mature Dentition

Two major groups of tooth morphologies were represented in our samples. One group (sharks, skates, tautog, gar, goldfish, cichlid, bluefish, and sea robin) had enameloid caps supported by tooth shaft dentin (Fig. 1), whereas the other group had enameloid pieces with little (ballonfish, and porcupine fish) or no dentin (parrotfish and sharpnose pufferfish) (Figs. 2, 3). Tautog, cichlid, and bluefish had thin layers of collar enameloid on their tooth shaft surface, whereas the gar displayed a hypermineralized layer of ganoid/enamel containing incremental lines [11, 12] not found on collar enameloid.

Fluoride Content of Enameloid

Electron microprobe analysis (by wavelength dispersive spectroscopy; WDS) showed a species-specific, high (2.0–3.7 wt%) or low (<0.4 wt%) fluoride content in the mature enameloid. Species containing high fluoride in enameloid were the shark, tautog, cichlid, sea robin, and parrotfish whereas pufferfish, bluefish, goldfish, and gar had low fluoride. The fluoride content of the gar ganoid/enamel layer was also low. All species had consistently low fluoride (0.8–0.1 wt%) in their dentin and bone. Energy dispersive spectroscopy (EDS) linescan analysis across the tooth-bone boundary indicated no fluoride gradient between pufferfish enameloid and bone (Fig. 4), whereas a sharp fluoride gradient was observed between parrotfish enameloid and bone (Fig. 5). Previously, we demonstrated that fluoride accumulates in developing enameloid matrix, and that the final fluoride content of the enameloid is not related to serum fluoride levels [9]. Thus, a specific fluoride-binding protein or cellular process must be responsible for concentrating fluoride into enameloid.

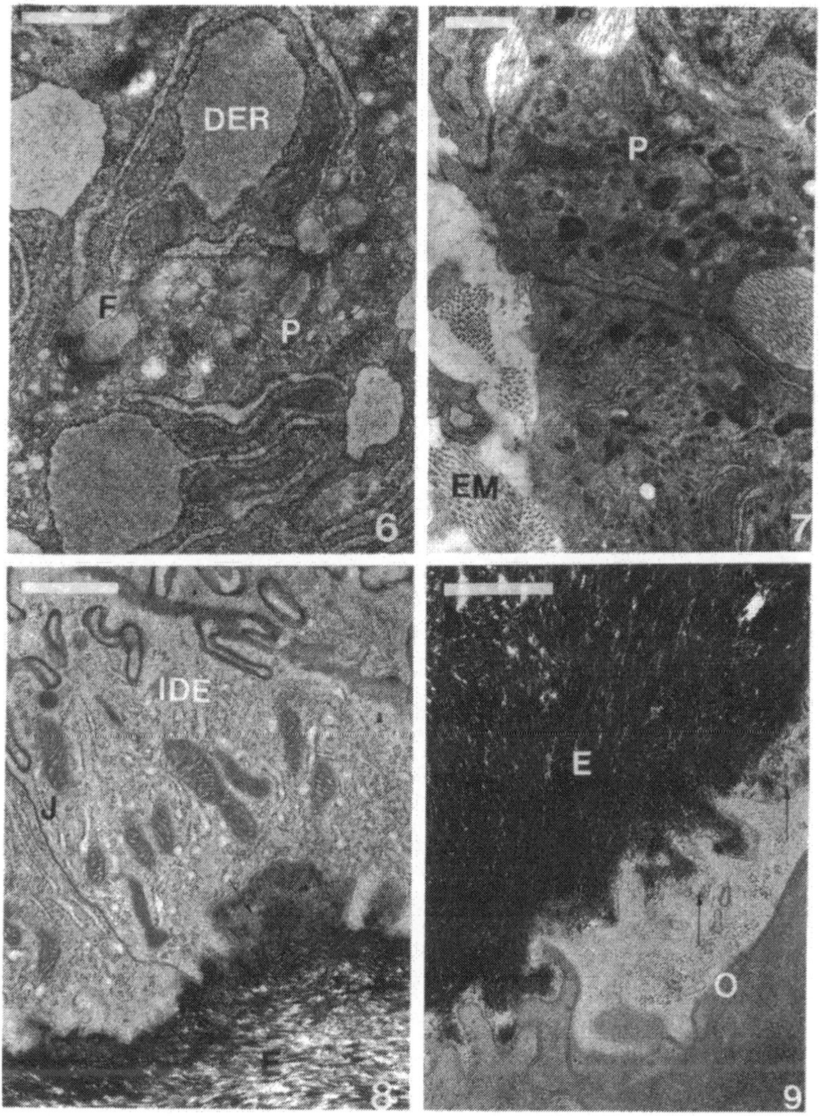

Fig. 6. TEM of colchicine-treated sharpnose pufferfish IDE cells at the enameloid matrix secretion stage. Dilated granular endoplasmic reticulum (*DER*), numerous procollagen granules (*P*), and striated fibers (*F*) were characteristic of colchicine-treated IDE cells. *Scale* = 1 µm

Fig. 7. TEM of colchicine-treated sharpnose pufferfish odontoblasts associated with enameloid matrix (*EM*). Accumulations of procollagen granules (*P*) were frequently but not consistently observed in all odontoblasts. *Scale* = 1 µm

Fig. 8. TEM of amorphous surface layer on mineralizing sharpnose pufferfish enameloid (*E*). The newly secreted material has small crystallites with random orientation (*arrows*). Intercellular junctions (*J*) are observed between neighboring *IDE* cells. *Scale* = 1 µm

Fig. 9. TEM of amorphous material (*arrows*) associated with sharpnose pufferfish odontoblasts (*O*) at the enameloid mineralization stage. *Scale* = 1 µm

Development of Enameloid Matrix

The presence of procollagen granules within both normal and colchicine-treated IDE cells from pufferfish (Fig. 6), parrotfish, cichlid, tautog, goldfish, and gar demonstrates that the bony fish enameloid matrix collagen is synthesized and secreted by the ectoderm. Procollagen granules have been seen in IDE cells of salmon [13] and eel (P. Shellis, personal communication, 1988).

It has been suggested that mesodermal collagen provides a scaffolding on which the ectodermal collagen is oriented [3]. However, the proportions of ectodermal to mesodermal collagens in enameloid remains unresolved [3, 13, 14]. Since parrotfish and sharpnose pufferfish contain no dentin (Figs. 2, 3), they are ideal models to investigate the relative contributions of the ectoderm and mesoderm to the enameloid matrix collagen.

Both pufferfish (Fig. 6) and parrotfish IDE cells responded to colchicine injection by accumulating procollagen granules in their cytoplasm at the enameloid matrix secretion stage. Puffer and parrotfish odontoblasts were confined to a small layer of cells at the base of each tooth bud. In colchicine-treated specimens, accumulations of procollagen granules were sometimes, but not consistently, observed within puffer and parrotfish odontoblasts (Fig. 7). The data suggest that the odontoblasts provide some "scaffolding" collagen which orients the ectodermally derived collagen into fibers [3]. Since this "scaffolding" collagen is present in both high and low fluoride-containing enameloid matrices, it is unlikely that the origin of collagen influences the fluoride content of enameloid.

Non-Enameloid Surface Layers

Pufferfish and parrotfish IDE cells contained non-collagenous, electron-dense secretory granules which were released onto the surface of the mineralizing enameloid (Fig. 8). Nascent crystallites subsequently formed in this material, but had no particular orientation and were quite different in size from the subjacent enameloid crystallites. At later stages of mineralization, pufferfish surface crystallites could not be distinguished from those of its enameloid, whereas parrotfish surface crystallites were somewhat shorter than those in the underlying enameloid.

Amorphous material was also observed between pufferfish odontoblasts and the mineralizing enameloid matrix (Fig. 9). This collagen-free odontoblast product also contained crystallites which, at later stages, were indistinguishable from enameloid crystallites. Possibly, odontoblasts can also express non-collagenous matrix proteins which modulate apatite formation during the mineralization stage.

Previous light and scanning electron microscopic studies suggested that in crossopterygian coelacanth, the IDE cells secrete a true enamel layer [15, 16]. An ectodermally derived enamel-like layer has also been described on lungfish (*Protopterus sp*) teeth [17]. However, the ultrastructure of the formation of these enamel-like layers has yet to be investigated. We have found that the

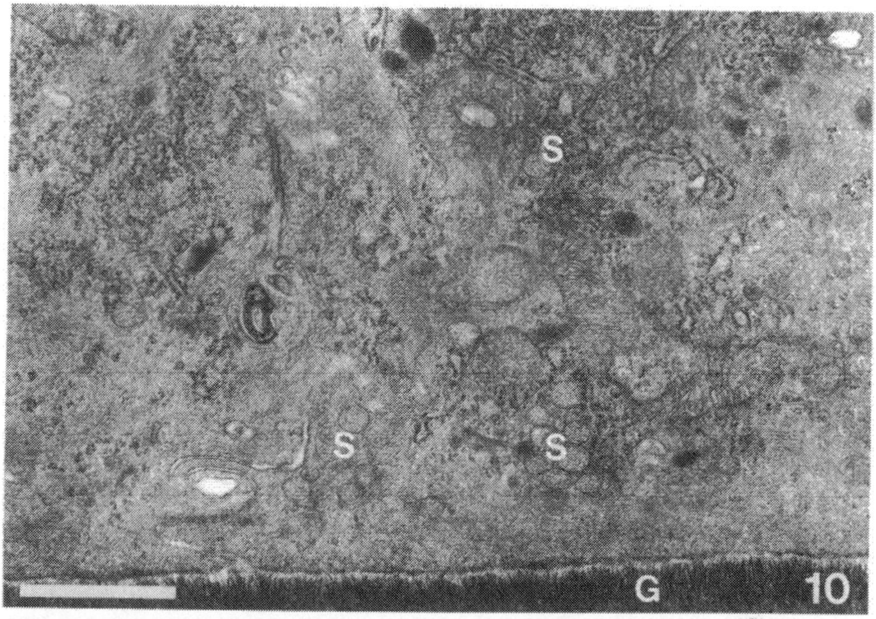

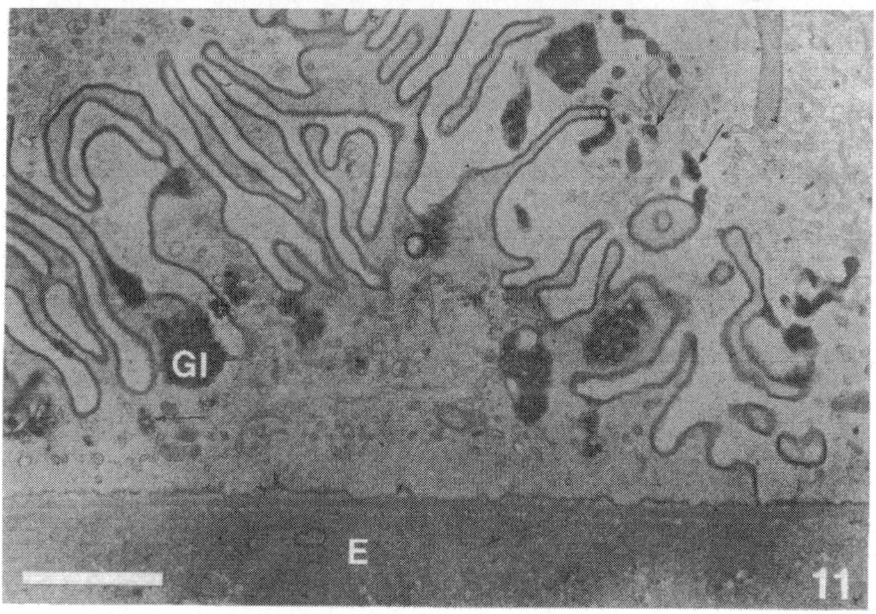

Fig. 10. TEM of ganoid/enamel formation in the alligator gar. Newly secreted enamel-like matrix (*G*) contains crystallites oriented normal to the IDE cell surface. Characteristic granules (*S*) contain "stippled" material. *Scale* = 1 μm

Fig. 11. TEM of glucagon-treated skate IDE cells, 1 h post-injection. As in colchicine-treated material, increased glycogen (*Gl*) is observed within intracellular vesicles (*arrow*) and between neighboring IDE cells. *Scale* = 1 μm tooth shaft of the alligator gar also contains a similar, non-collagenous, ectodermally derived enamel-like layer [11, 12].

The formation of ganoid/enamel is distinct from the parrotfish and pufferfish surface layers, which mineralize to become indistinct from the underlying enameloid. First, ganoid/enamel is found on the tooth shaft dentin and not on the enameloid crown itself [12] (Figs. 1, 10). Second, crystallites form immediately within the matrix and are oriented perpendicular to the tooth surface (Fig. 10). Third, characteristic secretory granules are observed within IDE cells associated with ganoid/enamel secretion but not within IDE cells associated with enameloid formation. The presence of incremental lines parallel to the tooth surface in ganoid/enamel bear a remarkable resemblance to mammalian aprismatic enamel, and supports its true ectodermal origin [11]. However, the biochemical properties of the various surface layers should be characterized to determine whether they do, in fact, represent a simple enamel.

Chondrichthian Enameloid

Chondrichthian enameloid has long been known to differ from osteichthian enameloid [18–21]. The ultrastructure of the matrix is unique in that the IDE cells do not synthesize large quantities of proteins during its formation, and the characteristic association of "giant" fibers with odontoblast processes strongly suggest their mesodermal and non-collagenous nature [8, 19]. The IDE cells store large quantities of glycogen [8, 18, 19, 21] which is secreted and then degraded in intercellular spaces [8, 19]. Our preliminary evidence using insulin and glucagon to interfere with glycogen metabolism in skates indicated that glucagon, but not insulin, produced effects similar to colchicine [19], causing increased exocytosis of glycogen-containing vesicles into intercellular spaces (Fig. 11). Glycogen (or its breakdown products) may possibly play a role in osmoregulation of the enameloid matrix, or provide a source of phosphate for growing apatite crystallites.

Tubular, membrane-like material in the elasmobranch enameloid matrix has frequently been observed and thought to arise from odontoblast processes [20]. The different enameloid matrix morphologies observed between skates [8, 19] and sharks [20] may be due to species differences, chemical processing, or temporal differences. We have observed similar tubules in well-fixed skate specimens, but mostly in samples which have been demineralized. Changes in the ionic composition of the enameloid fluids may affect the conformation of the matrix materials, leading to altered matrix morphologies. Additionally, the removal of material from the matrix by prolonged fixation may unmask the tubular structures. Future studies are being geared towards resolving these questions.

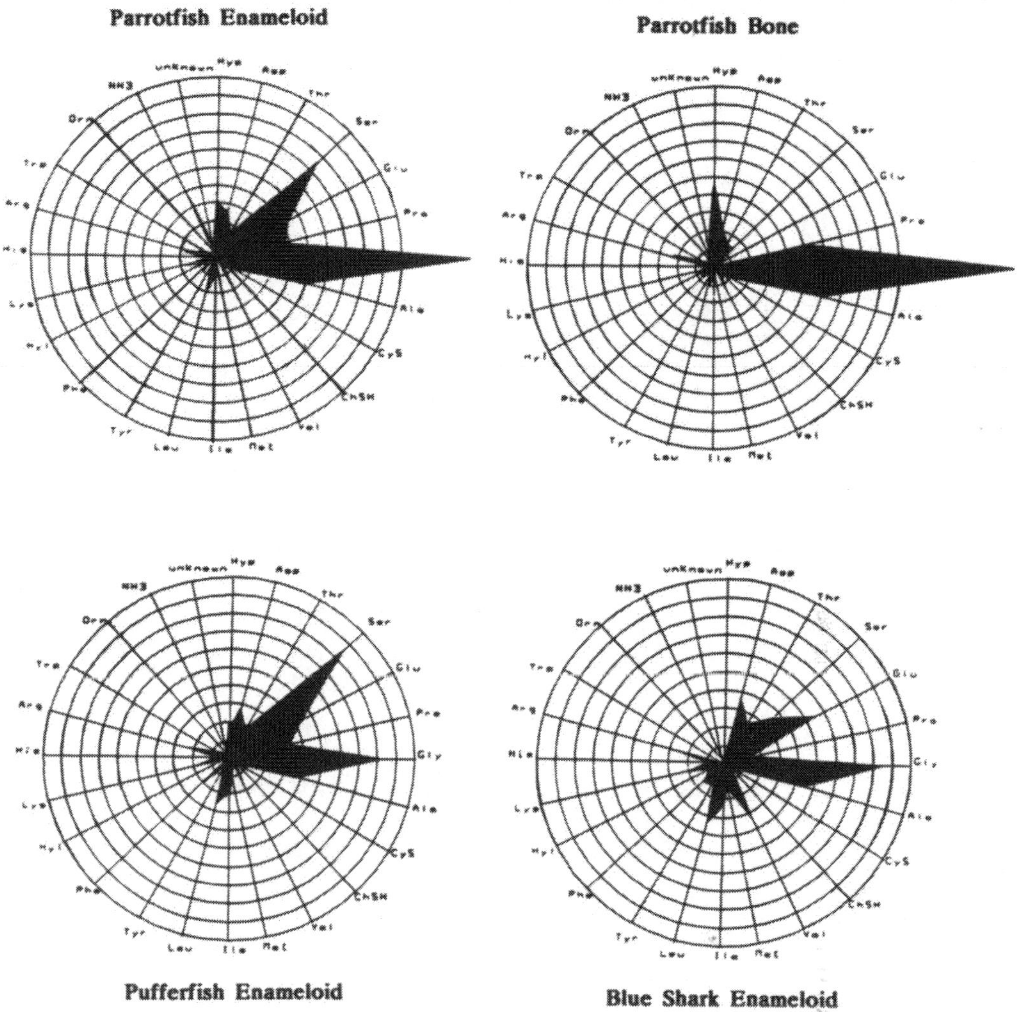

Fig. 12. Rose diagrams depicting the amino acid composition of parrotfish, pufferfish, and blue shark enameloid and parrotfish bone. Large collagen and acidic protein components are present in bony fish having both high (parrotfish) and low (pufferfish) fluoride in their enameloid. Blue shark enameloid, from which the collagen has been subtracted, contains high proportions of basic amino acids relative to bony fish enameloid proteins

Enameloid Matrix Biochemistry

Amino acid analyses of parrotfish and pufferfish enameloid matrices demonstrate the presence of collagen and other acidic proteins (Fig. 12). The bony fish enameloid matrix was enriched in acidic proteins compared to mature fish bone (Fig. 12) and had phosphorylated serine residues (100 res/1000) which

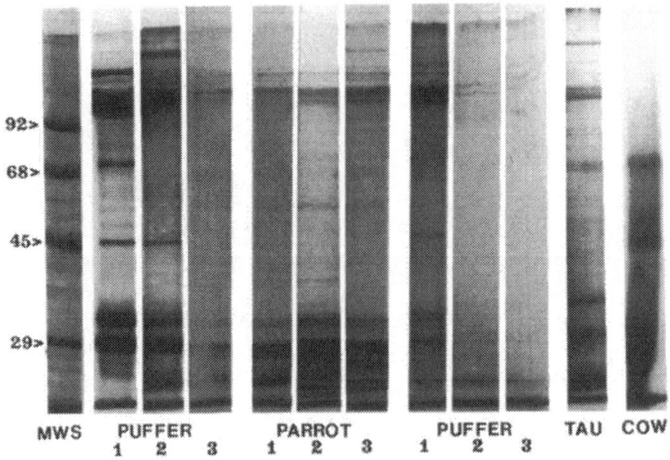
92>
68>
45>
29>
MWS
PUFFER
1 2 3
PARROT
1 2 3
PUFFER
1 2 3
TAU
COW

13

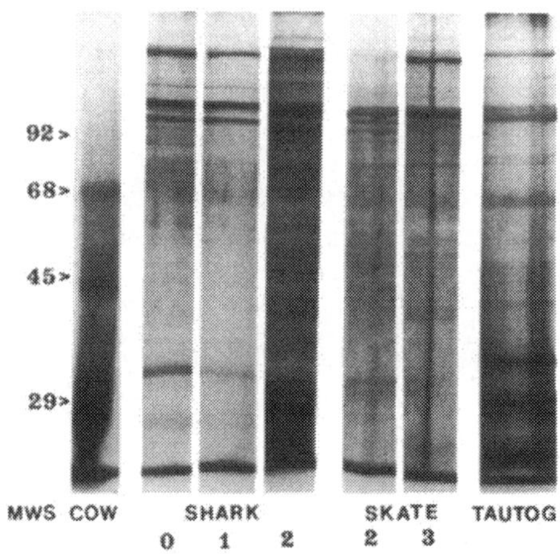
92>
68>
45>
29>
MWS
COW
SHARK
0 1 2
SKATE
2 3
TAUTOG

14

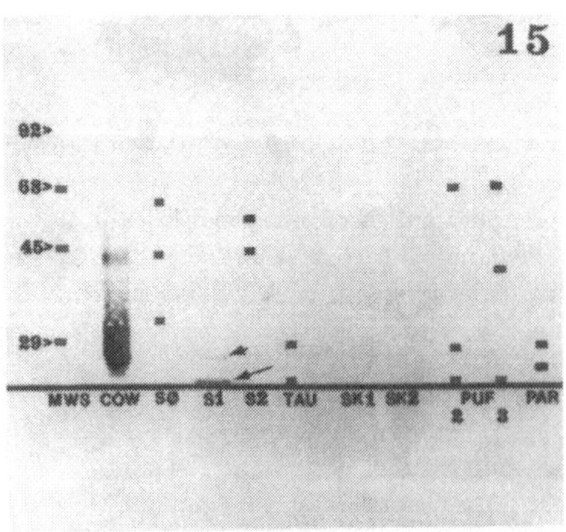
15
92>
68>
45>
29>
MWS COW S0 S1 S2 TAU SK1 SK2 PUF PAR
2 3

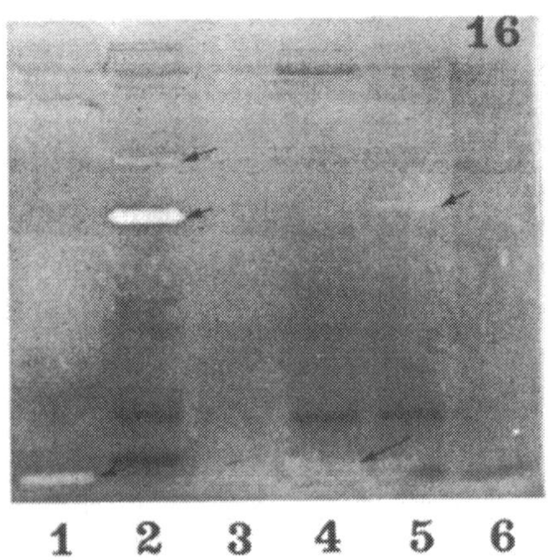
16
1 2 3 4 5 6

declined with increased mineralization (30 res/1000) in both fish enameloids. The overall amino acid compositions between the high (parrotfish) and low (pufferfish) fluoride enameloid matrices were strikingly similar. Thus, it is unlikely that a fluoride-binding protein accounts for the differences in fluoride content between the enameloids of these fish species.

The amino acid composition of the blue shark enameloid, corrected for collagen content, revealed a somewhat different profile from those of bony fish enameloid (Fig. 12). Enrichment of the basic amino acids in the shark enameloid matrix indicated a non-amelogenin type molecule [22], an admixture of actin from cytoplasmic processes, keratin-like proteins [8, 19], or contaminating serum albumin [22].

The electrophoretic patterns of acetic acid soluble proteins were quite similar between the high (parrotfish and tautog) and low (pufferfish) fluoride-containing enameloid matrices of bony fishes (Fig. 13), but different from the elasmobranchs (Fig. 14). All fish species showed characteristic Type I collagen doublets at 100 kd, a weak 68 kd band, but only bony fish enameloid had prominent 33 and 30 kd bands. With progressive mineralization (tooth rows 1, 2, and 3), the 33 and 30 kd bands decreased in intensity and whereas bands below 30 kd transiently appeared more intense. The electrophoretic fingerprints of bony fish enameloid strongly resembled peptide fragments from pollack bone [23] and are suggested to represent degradation peptides of the enameloid collagen itself. Proteases have been localized in enameloid [24], and have similar molecular weight (MW) between the high and low fluoride bony fish groups (Fig. 15). The extent of proteolysis, however, is not known.

Electrophoretic patterns of elasmobranch enameloid proteins were different from both bony fish enameloid and bovine enamel matrix proteins. Although elasmobranch enameloid matrix samples contained some dentin collagen, few

←

Fig. 13. SDS-PAGE of bony fish enameloid and cow enamel matrix proteins. Early stage matrices (*row 1*) have major bands at 100, 33, and 30 kd, whereas more mineralized matrices (*rows 2, 3*) have fewer high MW bands but more transient bands below 30 kd. All species have a 68 kd band, indicating albumin. *MWS*, Molecular weight standards; *TAU* = tautog

Fig. 14. SDS-PAGE of elasmobranch and tautog enameloid proteins. The 100 kd doublet in elasmobranch enameloid most likely represents dentin collagen. Note that the other bands, except for those at 100 and 68 kd, do not correspond with those of bony fish enameloid, in this case, tautog (high fluoride)

Fig. 15. Western blotting of cow enamel and fish enameloid proteins using rat enamel matrix antisera. Cow enamel is distinctly cross-reactive whereas tautog (*TAU*), pufferfish (*PUF*), parrotfish (*PAR*), and skate (*SK1*, *SK2*) enameloid proteins are unreactive. Only row 2 of blue shark (*S1*) had two cross-reactive proteins with this sera (*arrows*). *Rectangles* represent proteins which were positively transferred, identified by Coomassie blue staining

Fig. 16. Gelatin gel depicting gelatinase activity (*arrows*) in the low fluoride porcupine fish (*lane 1*, row 2; *lane 2*, row 1) and the high fluoride parrotfish (*lanes 3–5*, rows 4, 2, and 1) enameloid of the lower MW collagen degradation bands were observed (Fig. 14). Paucity of low MW bands reinforces the idea that elasmobranch enameloid matrix is primarily a keratin-like protein, different from bony fish enameloid collagen matrix [8, 19].

Homology

Previously, we had shown that antisera to mammalian enamel matrix proteins did not cross-react with bony fish enameloid matrix proteins [11]. Thus, the lower MW bands of bony fish enameloid are not related to the amelogenin class of enamel proteins. High resolution TEM immunocytochemical analyses of bony fish enameloid matrices, surface amorphous layers, and ganoid/enamel also produced negative results with this molecular probe [11]. However, we subsequently found that two blue shark enameloid matrix proteins cross-reacted with the enamel antisera (Fig. 16). However, our cross-reactive shark proteins had a lower MW (<20 kd) than the 66 kd protein reported in cat shark enameloid [25]. Possibly, mimeotopes may exist between structurally distinct proteins which are involved in calcification processes [26].

Other studies suggest that an enamel-like protein is present in fish enameloid [27]. If definitive evidence substantiates this hypothesis, then this protein can not be strongly related to the fluoride content of enameloid. Specifically, enamel-like antigens have been reported in both sharks of a high fluoride species, as well as in the low fluoride-containing enameloids of cod and bluefish [27]. Additionally, the mature surface layers in pufferfish (Fig. 4; low fluoride) and parrotfish (Fig. 5; high fluoride) have the same fluoride content as the underlying enameloid. Thus, even in areas enriched in non-collagenous proteins, the fluoride content of the enameloid was not affected.

Enameloid Collagen Cross-Links

The putative non-collagenous ectodermal proteins of enameloid have been suggested to interfere somehow with collagen cross-linking [2]. However, this hypothesis has not been substantiated. Two cross-link amino acids, dihydroxylysinonorleucine (DHLNL) and hydroxylysinonoleucine (HLNL), are commonly associated with bone and dentin collagens. Preliminary data (courtesy of Y. Kuboki, 1991) concerning these cross-links indicate several interesting features in fish enameloid and bone. In bone, the ratios of DHLNL/HLNL appear species specific [23], with HLNL cross-links reduced in fish compared to bovine bone. Generally, few differences were observed in the relative degree of cross-linking between high fluoride parrotfish developing enameloid and bone collagen (Fig. 17). Likewise, at similar stages of development, the collagen in low fluoride pufferfish enameloid had a nearly identical DHLNL/HLNL ratio to that in the high fluoride parrotfish enameloid (Fig. 17). Thus, the biochemical evidence strongly suggests that there is no interference in bony fish enameloid collagen cross-linking when compared to its homologous bone collagen. Likewise, the similarity in the other biochemical

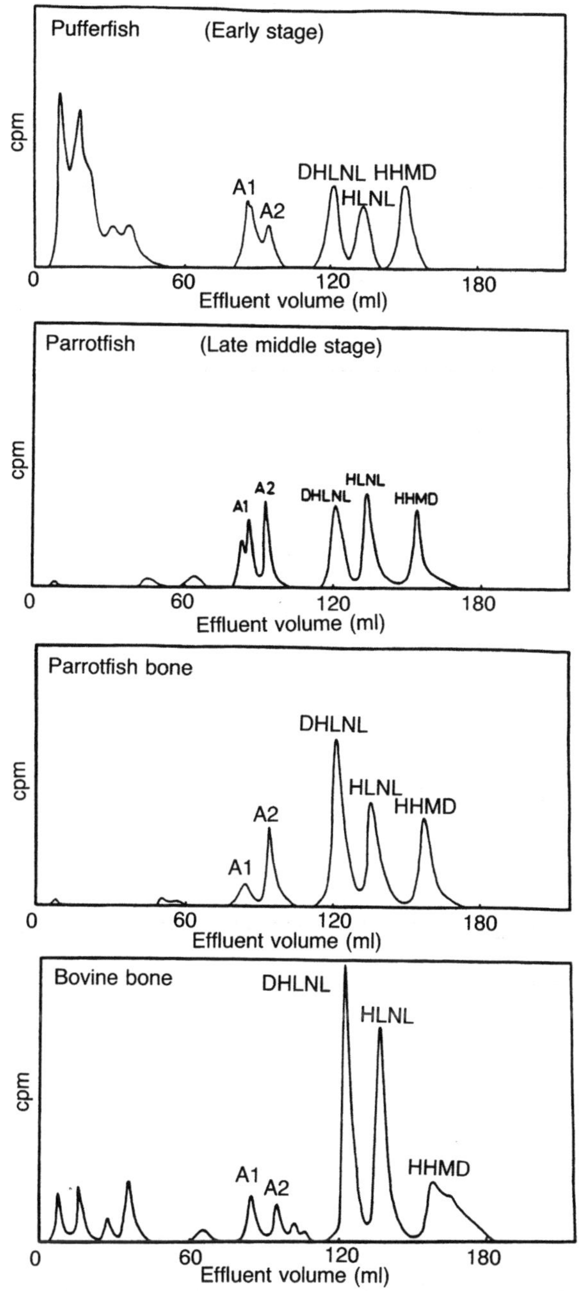

Fig. 17. Chromatograph demonstrating cross-link amino acids, dihydroxylysinonorleucine (*DHLNL*), hydroxylysinonoleucine (*HLNL*) and histidinohydroxymerodesmosine (*HHMD*), in developing parrotfish and pufferfish enameloid and in parrotfish and bovine bone. *A1* and *A2* peaks indicate the alpha 1 and 2 chains of collagen, respectively. The quantity of cross-links between both high and low fluoride enameloid and fish bone collagens are very similar

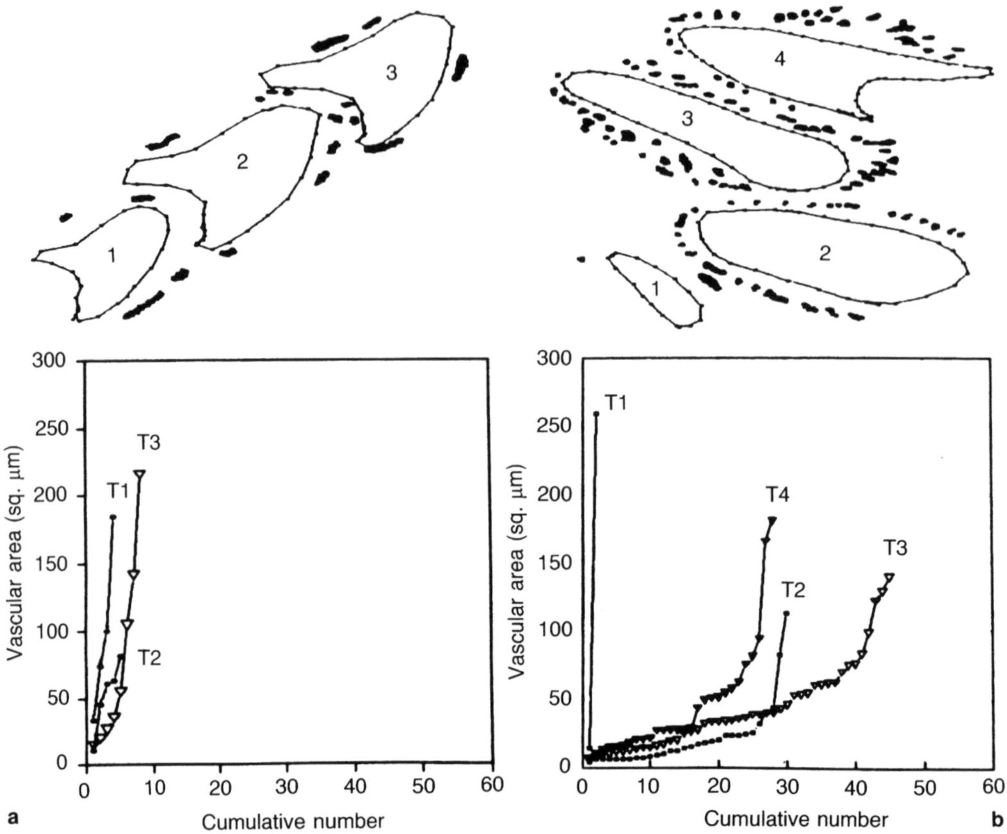

Fig. 18. Digitized images of light microscopic sections of pufferfish (**a** *rows 1–3*) and parrotfish (**b** *rows 1–4*) tooth buds. *Solid objects* represent capillaries surrounding the ODE cells. For each tooth bud and species, the vascular cross-sectional area is plotted against the cumulative number of vessels present. The high fluoride parrotfish displays numerous small capillaries throughout the mineralization stages (**b** *rows T2–4*), whereas the low fluoride, pufferfish capillaries are relatively few and large (**a**) with no change occuring during enameloid mineralization (*rows T2–3*)

features between the enameloid matrices of high and low fluoride bony fish makes it improbable that a fluoride-binding matrix protein can account for the difference in the final fluoride content of bony fish enameloid.

Interestingly, the quantity of DHLNL (0.93 mol/mol collagen) and HLNL (0.57 mol/mol collagen) cross-links in developing blue shark teeth (enameloid with adhering dentin) was approximately 3 times higher than bovine dentin (0.33 mol/mol collagen; 0.1 mol/mol collagen, respectively) (not shown). These data suggest that the collagen present in shark teeth is of odontoblastic origin. Secondly, the lack of interference with the cross-linking of the mixed enameloid-dentin sample supports the idea that chondrichthian enameloid is a

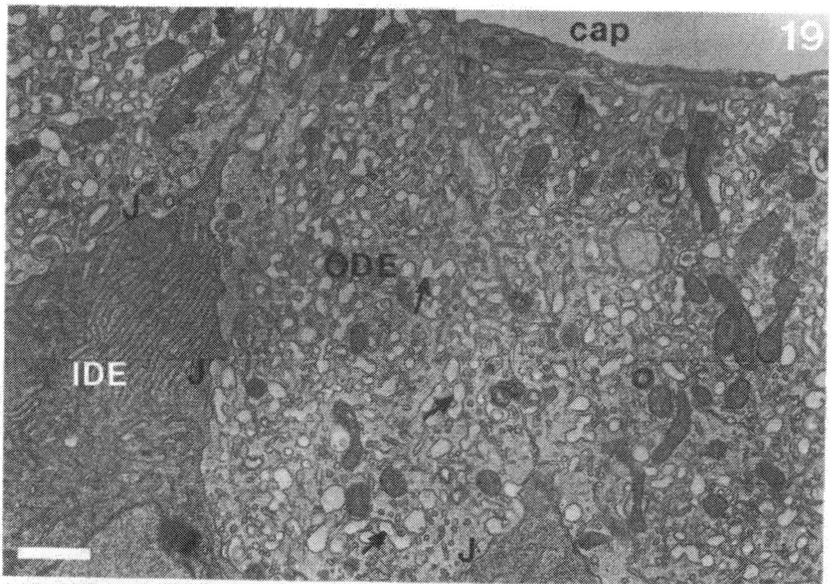

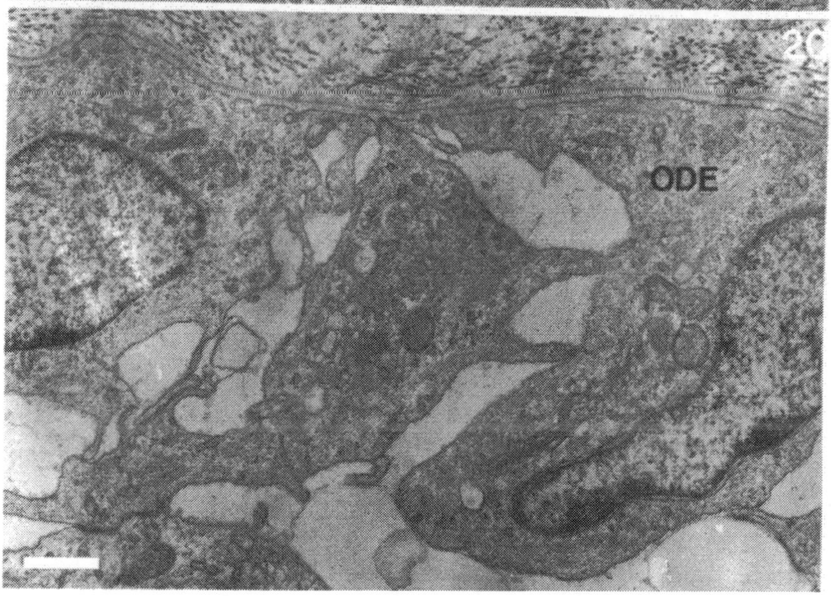

Fig. 19. High fluoride, cichlid ODE cells at the enameloid mineralization stage. Fenestrated capillaries (*cap*) are juxtaposed to the ODE basal lamina. A network of tubular structures (*arrows*), some contiguous with the plasma membrane, fill the cytoplasm of the ODE cells. Numerous mitochondria are present and extensive gap junctions (*J*) connect ODE with IDE cells. The specialized structure of ODE cells is remarkably similar to that seen in "chloride cells" of gills. *Scale* = 1 μm

Fig. 20. Low fluoride, alligator gar ODE cells at the enameloid mineralization stage. Capillaries are rarely found near the ODE cells in the low fluoride enameloid organs. ODE cells are not specialized, have few intercellular junctions, and display few tight junctions with neighboring IDE cells. *Scale* = 1 μm unique non-collagenous protein [8, 19, 20]. Possibly, this unique enameloid matrix does have fluoride-binding properties.

A Hypothesis of Fluoride Regulation During Enameloid Matrix Mineralization

The ionic composition and pH are important factors which influence the incorporation of fluoride into apatite in vitro [28]. However, our preliminary EDS data indicate that the concentrations of detected ions (Na, Cl, Ca, and PO_4) in developing enameloid are similar between high- and low-fluoride species. Additionally, high pH values (around 10.5) can decrease the fluoride content in forming apatite by fourfold [28], but are unlikely to occur in vivo [28] and would not account for the 37-fold difference in fluoride content between high and low fluoride species.

Instead, we suggest that a triad of cytostructural features, characteristic of enameloid organs in high fluoride-containing bony fish tooth buds, contributes to fluoride transport and concentration in enameloid. These features, present specifically during the enameloid mineralization stage, are: (1) an outer dental follicle which is highly vascularized with fenestrated capillaries (Fig. 18), (2) a highly invaginated, tubular network of membranes within ODE cells [3] (Fig. 19) which are remarkably similar to those observed in "chloride cells" of

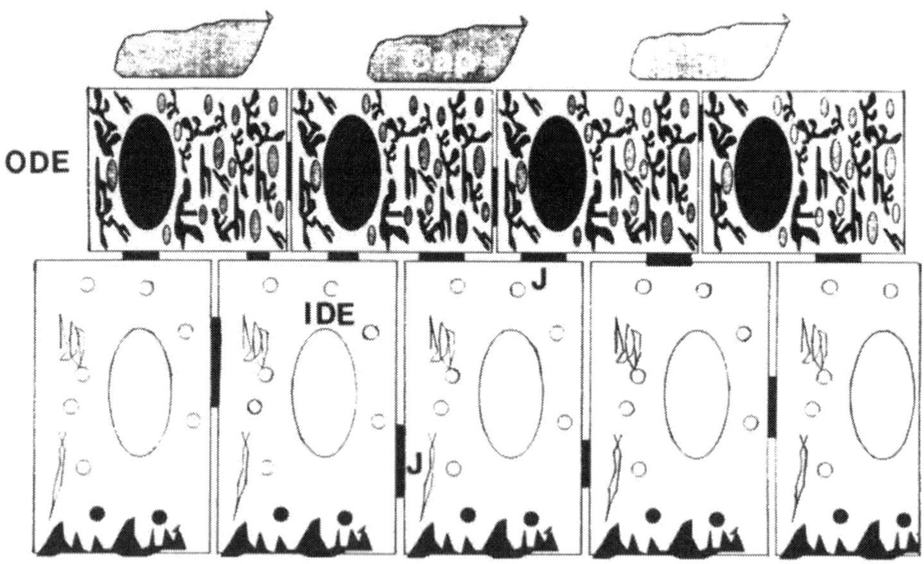

Fig. 21. Diagrammatic representation at a mineralization stage of a bony fish enameloid organ forming a high fluoride-containing enameloid. Numerous fenestrated capillaries (*Cap*) are closely juxtaposed to specialized ODE cells possessing a network of tubular membrane invaginations (*solid irregular structures*). Extensive gap junctions connect ODE to IDE cells (*heavy lines*)

fish gills [29], and (3) tight junctions or gap junctions between ODE and IDE cells which facilitate intracellular diffusion of ions towards the enameloid matrix while limiting intercellular diffusion [3]. High fluoride bony fish species which to date display all these features are cichlid, tautog, sea robin, and the Labrus and Sargus species. Low fluoride species which show neither a rich vasculature, specialized ODE cells, nor numerous intercellular junctions are the goldfish and alligator gar (Fig. 20). Although the low fluoride pufferfish have somewhat specialized ODE cells, their tubular networks are not extensive, nor are they highly vascularized. Thus, the combination of a low but continuous source of fluoride through fenestrated capillaries, specialized ODE cells, and numerous gap junctions are strongly related to the fluoride content of enameloid (Fig. 21).

We propose that the similarity in structure between the specialized ODE cells and "chloride cells" of fish gill epithelia is also reflected in their function. In "chloride cells", serum chloride enters the cell through sodium chloride symport channels [30]. Subsequently, large quantities of sodium are extruded from the "chloride cell" by the membrane Na-K ATPase, thus providing an

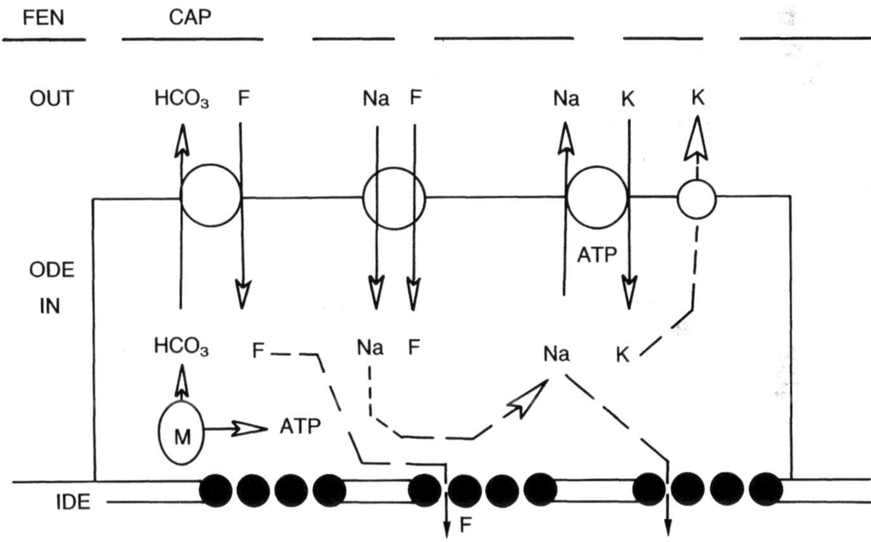

Fig. 22. Diagrammatic representation of fluoride transport through the specialized ODE cell. Fenestrated capillaries (*Fen Cap*) allow for the free diffusion of ions towards the ODE cell tubular invaginations. NaF symport channels allow these ions to preferentially diffuse into the ODE cytoplasm. The Na-K ATPase maintains the intracellular levels of Na low and K high. Another plausible mechanism for fluoride transport into ODE cells is via a bicarbonate-fluoride antiport system. Here, mitochondria release CO_2 resulting from ATP production, thus driving the bicarbonate-fluoride antiport system. Ionic fluoride, unable to diffuse through the plasma membrane, is driven down its concentration gradient through gap junctions connecting ODE to IDE cells towards the enameloid

electrochemical gradient to drive chloride into the environment [31]. In fact, "chloride cells" have the highest quantity of membrane Na-K ATPase of any tissue [31]. Since fluoride is the most electronegative ion and is a halide similar to chloride, it is reasonable that fluoride can be preferentially selected over chloride from the extracellular fluid (Fig. 22). Ionic selectivity is observed in cationic membrane pumps (Na/K ATPase) and through anionic membrane channels where chloride is preferentially secreted over fluoride [32, 33]. Presently, we are investigating the possibility that specialized ODE cells are also rich in membrane Na-K ATPase, and whether anion-selective channels are specific for fluoride. This hypothesis raises the possibility that the anion-selective membrane channels have a fluoride-"binding" property capable of distinguishing fluoride from chloride. Future studies are aimed at clarifying this proposed mechanism of fluoride transport towards the enameloid matrix.

Acknowledgements. This work was supported by NIDR Grant 07677. The authors express their gratitude to Dr. Yoshinori Kuboki and staff, Department of Biochemistry, Hokkaido University, Japan, for performing the amino acid composition and collagen cross-linking analyses of enameloid matrices.

References

1. Peyer B (1968) Comparative Odontology. Zangerl R (ed) University of Chicago Press, Chicago
2. Shellis P, Miles AEW (1974) Autoradiographic study of the formation of enameloid and dentine matrices in teleost fishes using tritiated amino acids. Proc R Soc Lond 185:51–72
3. Prostak K (1984) An ultrastructural study of enameloid matrix formation and mineralization in a teleost, *Cichlasoma cyanoguttatum*, using selected experimental systems. PhD Dissertation, Boston University, Boston, Mass.
4. Prostak K, Skobe Z (1985) The effects of colchicine on the ultrastructure of the dental epithelium and odontoblasts of teleost tooth buds. J Craniofac Genet Dev Biol 5:75–88
5. Suga S, Wada K, Ogawa M (1980) Fluoride concentration in the enameloid of fishes. In: Omori M Watabe N (eds) The Mechanisms of biomineralization in animals and plants. Tokai University Press, Tokyo, pp 229–240
6. Suga S, Wada K, Ogawa M (1981) Fluoride concentration in teeth of tetraodontiform fishes and its phylogenetic significance. Jpn J Ichthyol 28:304–312
7. Suga S, Taki Y, Wada K, Ogawa M (1991) Evolution of fluoride and iron concentrations in the enameloid of fish teeth, In: Suga S, Nakahara H (eds) Mechanisms and phylogeny of mineralization in biological systems. Springer, Tokyo, pp 439–446
8. Prostak K, Seifert P, Skobe Z (1990) The effects of colchicine on the ultrastructure of odontogenic cells in the common skate, *Raja erinacae*. Am J Anat 189:77–91
9. Prostak K, Seifert P, Skobe Z (1991) Serum fluoride level and fluoride content of enameloid. J Dent Res 70(9):1266–1271
10. Kuboki Y, Tsuzaki M, Sasaki S, Liu CF, Mechanic G (1981) Location of the intermolecular cross-links in bovine dentin collagen, solubilization with trypsin

and isolation of cross-link peptides containing dihydroxylysinonorleucine and pyridinoline. Biochem Biophys Res Comm 102:119–126
11. Prostak K, Seifert P, Skobe Z (1991) Tooth matrix formation and mineralization in extent fishes. In: Suga S, Nakahara H (eds), Mechanisms and phylogeny of mineralization in biological systems. Springer, Tokyo, pp 465–469
12. Prostak K, Seifert P, Skobe Z (1989) Ultrastructure of developing teeth in the gar pike, (*Lepisosteus*). In: Fearnhead R (ed) Tooth enamel V, Florence, Yokohama, Japan
13. Sasagawa I (1984) Formation of cap enameloid in the jaw teeth of dog salmon, *Oncorhynchus keta*. Jpn J Oral Biol 26:477–495
14. Sasagawa I, Ishiyama M (1988) The structure and development of the collar enameloid in two teleost fishes, *Halichoreses poecilopterus* and *Pagrus major*. Anat Embryol 178:499–511
15. Miller W (1969) Tooth enamel of *Latimeria chalumnae* (Smith). Nature 221:1244
16. Smith M (1978) Enamel in the oral teeth of *Latimeria chalumnae* (Pisces: Actinistia): A scanning electron microscope study. J Zool Lond 185:355–369
17. Ishiyama M, Ogawa T (1983) Existence of true enamel on tooth plate in the lungfish, *Lepidosiren paradoxa*. Acta Anat Nippon 58:157–161
18. Tomes CS (1898) Upon the structure and development of the enamel of elasmobranch fishes. Philos Trans R Soc Lond [Biol] 190:443–466
19. Prostak K, Skobe Z (1988) Ultrastructure of odontogenic cells during enameloid matrix synthesis in tooth buds from an elasmobranch, *Raja erinacae*. Am J Anat 182:59–72
20. Sasagawa I (1989) The fine structure of initial mineralization during tooth development in the gummy shark, *Mustelus manazo*, Elasmobranchia. J Anat 164:175–187
21. Kerr T (1955) Development and structure of the teeth in the dogfishes, *Squalus acanthias*, L. and *Scyliorhynus caniculus* (L.). Proc Zool Soc Lond 125:95–114
22. Robinson C, Kirkham, Fincham A (1989) The enamelin/non-amelogenin problem. A brief review. In: Glimcher M, Lian J (eds) The chemistry and biology of mineralized tissues. Gordon and Breach, New York, pp 719–726
23. Kuboki Y, Watanabe T, Tazaki M, Takita H (1991) Comparative biochemistry of bone matrix proteins in bovine and fish. In: Suga S, Nakahara H (eds) Mechanisms and phylogeny of mineralization in biological systems. Springer, Tokyo, pp 495–499
24. Kawasaki K, Shimoda S, Fukae M (1987) Histological and biochemical observations of developing enameloid of sea bream. Adv Dent Res 1:191–195
25. Deutsch D, Palmon A, Defni L, Shenkman A, Sherman J, Fisher L, Termine J, Young M (1991) Enamelin and enameloid. In: Suga S, Nakahara H (eds) Mechanisms and phylogeny of mineralization in biological tissues. Springer, Tokyo, pp 73–77
26. Kakei M, Nakahara H, Takeyama H (1991) Immunoblotting study of developing enamel matrix proteins. In: Suga S, Nakahara H (eds) Mechanisms and phylogeny of mineralization in biological tissues. Springer, Tokyo, pp 95–99
27. Herold R, Rosenbloom J, Granovsky M (1989) Phylogenetic distribution of enamel proteins: Immunohistochemical localization with monoclonal antibodies indicates the evolutionary appearance of enamelins prior to amelogenins. Calcif Tissue Int 45:88–94
28. Shimoda S, Aoba T, Moreno E, Miake Y (1990) Effects of solution composition on morphological and structural features of carbonated calcium apatites. J Dent Res 69:1731–1740

29. Pisam M, Prunet P, Boeuf G, Rambourg A (1988) Ultrastructural features of chloride cells in the gill epithelium of the atlantic salmon, *Salmo slar*, and their modifications during smoltification. Am J Anat 183:235–244
30. Potts W (1977) Fish Gills. In: Gupta B, Moreton R, Oschman J, Wall B (eds) Transport of ions and water in animals. Academic, New York, pp 453–480
31. Rankin J, Davenport J (1981) Movement between fresh water and sea water. In: Rankin J, Davenport J (eds) Animal osmoregulation. Wiley, Chichester, pp 83–100
32. Hanrahan J, Alles W, Lewis S (1985) Single anion-selective channels in basolateral membrane of mammalian tight epithelium. Proc Natl Acad Sci 82:7791–7795
33. Valdivia H, Dubinsky W, Coronado R (1988) Reconstruction and phosphorylation of chloride channels from airway epithelium membranes. Science 242:1441–1444

Amelogenin Hydrolysis in Isolated Enamel

Ioanna S. Iatridi and Miles A. Crenshaw*[1]

Summary. The purpose of this research was to determine if amelogenin hydrolysis in enamel can occur in the absence of an enamel organ and to determine whether protease inhibitors or fluoride have an effect on the hydrolysis. Fetal bovine secretory stage enamel was incubated in a buffer alone or in the same buffer containing either a serine protease inhibitor, a sulfhydryl protease inhibitor, or sodium fluoride. Hydrolysis was assessed by quantitative protein determinations and by gel electrophoresis. The results indicated that about 25% of the amelogenins were hydrolyzed within 5 days at 37°C, and that this hydrolysis was reduced by serine and sulfhydryl protease inhibitors. Fluoride at 1 mM had no apparent effect on the degradation of amelogenins but appeared to reduce their release from the enamel.

Key words: Enamel—Amelogenins—Proteases—Fluoride

Introduction

During the secretory stage of enamel development, ameloblasts deposit a protein matrix that is mineralized to a limited extent. The proteins consist mainly of hydrophobic, proline-rich amelogenins with a smaller fraction of acidic enamelins [1, 2]. A gradual loss of the amelogenins appears to begin within hours of their secretion [3]. In the maturation stage, the remaining amelogenins are rapidly lost while the enamelins are partially conserved [1, 2, 4]. During this stage, additional mineral is deposited until the mature enamel is about 95% mineral and 2% matrix by weight.

The loss of amelogenins appears to result from proteolytic activity present in developing enamel which was first demonstrated by Suga [5]. The mechanism involves active hydrolysis of a parent amelogenin into smaller polypeptides by proteases [4]. These enzymes probably are derived from the ameloblasts [1, 5–10].

[1] Department of Pediatric Dentistry and Dental Research Center, University of North Carolina, Chapel Hill, NC 27599–7455, USA
* Corresponding author

It has been estimated by electrophoresis that the parent amelogenin is 25–30 kd and that, during enamel development, it is cleaved in steps toward a <10 kd end product [4, 7]. These small polypeptides are more soluble and mobile than their parent amelogenin and are able to migrate out of the forming enamel. If this hypothesis is correct, protein hydrolysis should take place in secretory enamel in the absence of ameloblasts and without further mineral deposition. Preliminary observations by Robinson and Kirkham [2] indicate that the breakdown of amelogenins does continue after the enamel organ is removed from rat incisors.

Protease activities, which have been identified in developing enamel by various investigators, are mainly attributed to serine proteases [6–10]. Carter et al. [10] isolated and purified a serine protease from developing enamel and showed its ability to hydrolyze enamel proteins. Recently, the presence of metalloproteinase activity has been reported. Gelatin was used to demonstrate this activity by Overall and Limeback [11]. DenBesten and Heffernan [12] used the 27 kd amelogenin, purified from secretory bovine enamel, as the substrate for these kinds of proteases. They concluded that metalloproteinases have relatively weak activity against amelogenin.

Fluorosed enamel has a higher protein content than normal enamel [12, 14–16]. DenBesten and Crenshaw [16] found that the increased protein content was first evident in the maturation stage of incisor enamel from rats that were given fluorotic levels of fluoride. They suggested that a reduction or delay in amelogenin loss from enamel may be the initial effect of fluorosis in developing enamel. Because an enamel protease was found to be indirectly inhibited by fluoride [8], it was suggested that this inhibition may account for the higher protein content in maturing enamel. Subsequently, DenBesten [17] showed that in developing fluorosed enamel there was, first, a delay in the hydrolysis and, second, a delay in the loss of amelogenins from the enamel in vivo. However, it is not known whether this effect can occur in the absence of ameloblasts. The demonstration of an inhibition of amelogenin hydrolysis by fluoride in isolated enamel would support the mechanism proposed by DenBesten and Crenshaw [16]. This would provide additional insight as to whether fluoride can produce fluorosis by affecting the enamel directly.

Materials and Methods

Enamel

Near-term fetal calves were obtained from freshly slaughtered cows at a local abattoir. The mandibles were quickly removed, packed in ice, and returned to the laboratory. Each mandibular tooth was dissected free, and the soft tissues covering the enamel carefully wiped away with cotton sponges. While it was still moist, the soft secretory-stage enamel apical to the white opaque boundary was removed, using a blunt spatula. A thin layer of enamel was left over the dentin to avoid contamination with dentin.

Incubation of Enamel

Hydrolysis of proteins in isolated enamel and inhibition of this hydrolysis was observed by the following method. Samples of 100 mg from pooled collections of secretory stage enamel were placed into separate plastic microcentrifuge tubes. A 1-ml volume of 50 mM tris acetate buffer, pH 7.40, was added to each enamel sample with or without the addition of a protease inhibitor or fluoride. The protease inhibitors and concentrations that were used were: 80 µg/ml aprotinin (used primarily as a serine protease inhibitor) [18], 2 mM phenylmercuric acetate (PMA) and 100 mM N-ethylmaleimide (NEM) (sulfhydryl protease inhibitors) [19, 20]. Preliminary experiments using the serine protease inhibitor, phenylmethylsulfonyl fluoride (PMSF), had no significant effect on the release of amelogenins into the incubation buffer. A separate study, conducted to determine whether or not the presence of this mineral affects the inhibitory activity of PMSF, indicated that this inhibitor was too unstable for these experiments [21]. The fluoride samples contained 1 mM sodium fluoride (NaF).

After the addition of the incubation solutions, the enamel was dispersed with a glass rod and placed on a rotating wheel at 37°C for 5 days. Other samples, incubated for the same period of time (without the addition of inhibitors or fluoride) at −20°C, served as controls (frozen controls). It was anticipated that the primary, high-molecular weight amelogenin would be hydrolyzed. The low molecular weight amelogenins produced by this hydrolysis would subsequently dissolve from the enamel into the tris acetate buffer.

After the incubation, each sample was centrifuged, and the supernatant was collected and stored at −20°C (tris extracts). The enamel pellet was dispersed and extracted overnight by stirring at 4°C with 1 ml 4 M guanidine hydrochloride (GuHCl) in tris acetate buffer, pH 7.40. This procedure extracted the amelogenin fraction that was not released into the tris buffer during the incubation [1]. The samples were centrifuged and the supernatants were collected and then stored at −20°C (GuHCl extracts).

Quantitative Protein Determination

Two protein assays were used to determine (1) the amount of protein and (2) the number of free amino groups in each extract. The bicinchromic acid protein assay (BCA) [22] was used to determine the amount of protein. Bovine serum albumin was used to construct the standard curve. The *o*-phthalaldehyde (OPA) assay was used to determine the number of free amino groups [23]. Leucine was used to construct the standard curve. An increase in the ratio of free amino groups to the amount of protein in the tris extracts, compared to the frozen controls, would indicate hydrolysis. A low ratio in the tris extracts incubated with inhibitors or fluoride, compared to the tris extracts incubated alone, would indicate inhibition of the hydrolysis.

Gel Electrophoresis

After completion of the quantitative protein determinations, all samples were dialyzed (3500 mw cut-off) and lyophilized. SDS polyacrylamide gel electrophoresis (SDS-PAGE) was subsequently used to separate the amelogenins and assess their hydrolysis [2–4]. The method described by Laemmli [24] was followed, using 13% polyacrylamide gels. Approximately 20 μg protein was applied in each well. The molecular weight standards used were: carbonic anhydrase (30 kd), soybean trypsin inhibitor (20.1 kd) and α-lactalbumin (14.4 kd). Peptides smaller than approximately 10 kd were not detectable in this electrophoresis system. The gels were fixed with 1% glutaraldehyde for 1 hour and then stained with Coomassie brilliant blue. Decreases in the relative abundance of high molecular weight amelogenins in the tris extracts incubated without inhibitors or fluoride, when compared to the frozen controls, would indicate hydrolysis. In the samples incubated with inhibitors or fluoride, the presence of amelogenins with molecular weights higher than those of the samples incubated with tris alone would indicate reduction of hydrolysis.

Statistical Analysis

Statistical significance of the quantitative differences between the means of samples was evaluated by using the Student's *t*-test. Values of *t* indicating a probability of less than 0.05 were regarded as statistically significant [25].

Results

Amelogenin Hydrolysis in the Absence of Enamel Organ

After the incubation of the samples containing no inhibitors or fluoride, approximately 25% of the amelogenins were extracted by the tris acetate buffer. Compared to a 2% release of amelogenins in the frozen controls, this difference was significant ($P < 0.05$) (Table 1).

The ratio of free amino groups to the amount of protein in the tris extracts was significantly greater, $P < 0.005$, in the incubated samples compared to the

Table 1. Percentage of total protein in tris extract

	n	*Mean*	*S.D.*	*Compared to tris control*
Frozen control	8	2.00	1.62	
Tris control	9	12.7	3.2	$P < 0.001$*
Aprotinin	5	12.6	5.1	$P < 0.5$
PMA	4	11.2	4.5	$P > 0.5$
NEM	4	14.6	4.5	$P > 0.5$
NaF	5	6.66	2.81	$P < 0.005$

* Compared to frozen control

Table 2. Moles free amino groups/gm protein in tris extract

	n	*Mean*	*S.D.*	*Compared to tris control*
Frozen control	8	0.478	0.220	
Tris control	9	2.156	1.196	$P < 0.001$*
Aprotinin	5	0.153	0.066	$P < 0.001$
PMA	3	0.987	0.317	$P < 0.025$
NEM	4	0.179	0.039	$P < 0.0001$
NaF	5	1.825	0.621	$P < 0.2$

* Compared to frozen control

Table 3. Moles free amino groups/gm protein in GuHCl extract

	n	*Mean*	*S.D.*	*Compared to tris control*
Frozen control	8	0.028	0.020	
Tris control	9	0.036	0.024	$P < 0.4$*
Aprotinin	5	0.031	0.020	$P < 0.5$
PMA	3	0.028	0.013	$P < 0.4$
NEM	4	0.018	0.013	$P < 0.3$
NaF	5	0.015	0.010	$P > 0.4$

* Compared to frozen control

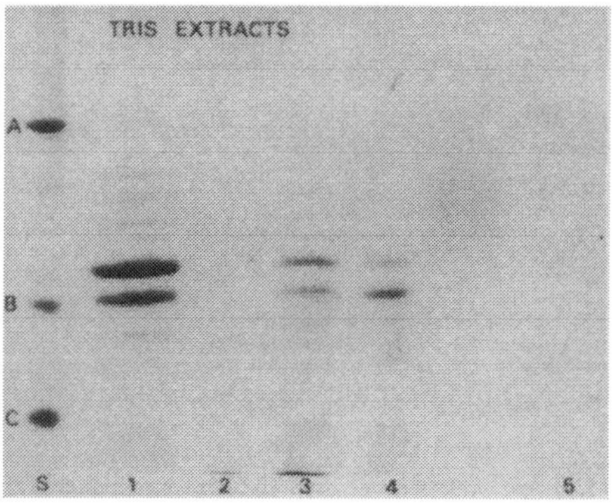

Fig. 1. SDS-PAGE of Tris Extracts. *Lane S*, Molecular size standards (A = 30 kd, B = 20.1 kd, C = 14.4 kd); *lane 1*, frozen control; *lane 2*, tris control; *lane 3*, aprotinin; *lane 4*, PMA; *lane 5*, NaF

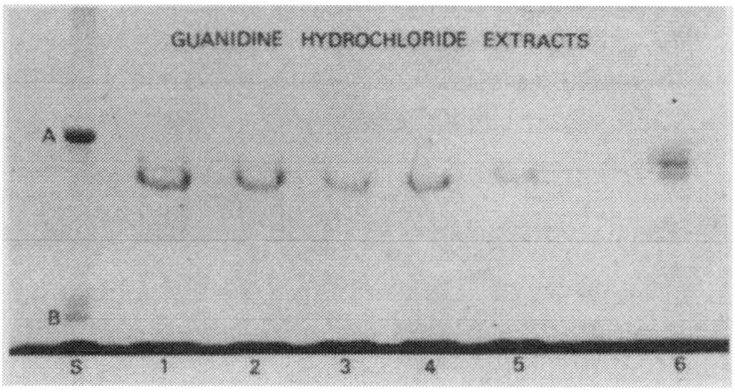

Fig. 2. SDS-PAGE of GuHCl Extracts. *Lane S*, Molecular size standards (A = 30 kd, B = 20.1 kd); *lane 1*, frozen control; *lane 2*, tris control; *lane 3*, aprotinin; *lane 4*, PMA; *lane 5*, NaF

frozen controls (Table 2). There were no differences in the amount of protein or in the ratio of free amino groups to the amount of protein between the guanidine-extracted controls and the guanidine-extracted experimentals (Table 3).

The electrophoretograms of the tris extracts showed that the 18 kd and 22 kd amelogenins were the most abundant proteins in the frozen controls. In the incubated tris extracts, no amelogenins greater than 10 kd were detected (Fig. 1). The guanidine extracted controls and experimentals were essentially identical, with a predominance of the 27 kd amelogenin and lesser amounts of the 28 kd amelogenin. No smaller amelogenins were detected (Fig. 2).

Effect of Protease Inhibitors on Amelogenin Hydrolysis

Sulfhydryl Inhibitors

Neither phenylmercuric acetate (PMA) nor N-ethylmaleimide (NEM) significantly inhibited the release of amelogenins into the medium (Table 1).

The ratio of free amino groups to the amount of protein in the tris extracts was significantly less for both PMA and NEM samples, $P < 0.05$ and $P < 0.005$, respectively (Table 2). In the guanidine extracts, this ratio was not significantly different between the samples containing either inhibitor and the incubated control (Table 3).

The electrophoretograms of the PMA and NEM tris extracts showed that the predominant amelogenins were 18 kd and 22 kd in these samples (Fig. 1). The GuHCl extracts of these samples were identical to the GuHCl extracts of the frozen controls and the GuHCl extracted samples containing no inhibitors or fluoride (Fig. 2).

Serine Protease Inhibitor

The amount of protein that was released into the tris acetate medium for the samples incubated with 80 μg/ml aprotinin was not significantly different from the samples incubated with tris alone (Table 1).

The ratio of free amino groups to the amount of protein in the tris extracts of these samples was significantly less at the $P < 0.001$ level when compared to the samples incubated in tris alone (Table 2). The difference in the ratios was not significant for the GuHCl extracts of these samples (Table 3).

The SDS-PAGE revealed that in the tris extracts of the aprotinin samples there was a predominance of the 18 kd and 22 kd amelogenins (Fig. 1). The guanidine extracts of the inhibited and control samples were identical (Fig. 2).

Effect of Fluoride on Amelogenin Hydrolysis

The inclusion of 1 mM NaF in the incubation buffer significantly diminished the amount of the amelogenin released into this medium, $P < 0.05$. Approximately 6% of the amelogenins were detected in this medium (Table 1).

The ratio of free amino groups to amount of protein was not significant for either the tris or the GuHCl extracts of the experimental samples when compared to their respective controls (Tables 2, 3).

No amelogenins above the detection limit of approximately 10 kd were observed in the tris extracts by SDS-PAGE (Fig. 1). The GuHCl extracts of the experimental and control samples were identical (Fig. 2).

Discussion

The results show that release of amelogenins from the enamel into the incubation buffer occurs during incubation of isolated secretory enamel at 37°C for 5 days. Approximately 25% of amelogenins were released from the developing enamel of the incubated samples. This release was significant when compared to the frozen controls. The hydrolysis of the released amelogenins was of primary interest since it is generally thought that the amelogenins are lost from the enamel after degradation of the primary 28.4 kd amelogenin into smaller polypeptides [4]. Our results (quantitative protein determination and SDS-PAGE) support this hypothesis. No amelogenins above 10 kd were detected in the tris extracts of the incubated enamel.

It is widely accepted that the primary amelogenins are hydrolyzed by proteases that are secreted from the ameloblasts together with the enamel matrix [5–9, 11, 12]. Our results clearly show that, indeed, amelogenins can be hydrolyzed in the absence of enamel organ and without further mineral deposition.

Several proteolytic activities have been identified in developing enamel by various investigators. These include serine proteases [6–10] and metalloproteinases [11, 12]. The roles of these enzymes in amelogenin loss has not been proven in vitro even though some have been shown to hydrolyze amelogenins [6, 9, 12]. Our results indicate that serine protease inhibitors are capable of reducing the hydrolysis of the amelogenins once they are released into the tris medium. The effect of the sulfhydryl protease inhibitors may indicate that this enzyme activity is responsible, in part, for amelogenin hydrolysis. However, sulfhydryl protease inhibitors may also inhibit some serine proteases by disrupting disulfide bonds necessary for the activity of these enzymes [9, 10]. Neither serine nor sulfhydryl protease inhibitors reduced the release of amelogenins from the enamel.

The model system for the study of amelogenin hydrolysis used in this research had a large fluid to enamel volume ratio, opposite from the ratio in vivo. For this reason, the model allowed the study of intermediate stages of amelogenin hydrolysis. The 28.4 kd primary amelogenin was found only in the guanidine extracts of the enamel. This observation confirmed that this protein has a significant affinity for the mineral. The partial hydrolysis of this protein to the 18–21 kd amelogenins eliminated this affinity, and this group of proteins was released from the mineral into the tris buffer. This step was evident in the tris extract containing inhibitors. In the absence of inhibitors, the 18–21 kd amelogenins were hydrolyzed into peptides too small to be detected by the electrophoretic system used.

The first step in amelogenin hydrolysis, the production of the 18–21 kd amelogenins, is the rate-limiting step. It appears that this step is specific for amelogenin hydrolysis. None of the protease inhibitors used in this study affected this first step. The proteolytic activity could be reduced only after the release of the amelogenins into the incubation medium. Therefore, we cannot conclude that the serine and sulfhydryl protease activities observed in this study are necessarily unique in the developing secretory enamel; any number of enzymes may digest the amelogenins after their release into the medium.

Suga [5] demonstrated inhibition of proteolytic activity in enamel by 5 mM NaF. Crenshaw and Bawden [8] found that the serine protease activity present in decalcified embryonic bovine secretory enamel could be inhibited indirectly by 5 μM fluoride at low calcium concentrations. In the current study of developing enamel, we elected to use a high NaF concentration, 1 mM. We reasoned that fluoride at this concentration would dramatically affect amelogenin loss from the enamel. The results indicated that fluoride did not inhibit the hydrolysis of amelogenins released into the tris medium.

Fluoride at this concentration did reduce the release of amelogenins from the enamel. It has been shown that the delay in the loss of amelogenins from fluorosed enamel [16] may be due to a direct effect of the fluoride on the enamel. Fluoride may have altered the enamel crystal. This would make the adsorbed 28.4 kd amelogenin unavailable for the primary hydrolysis. Alternatively, fluoride may inhibit this primary hydrolysis.

Acknowledgements. This work was supported, in part, by USPHS grant #DE05504.

References

1. Termine JD, Belcourt AB, Christner PJ, Conn KM, Nylen MU (1980) Properties of dissosiatively extracted fetal tooth matrix proteins. I. Principal molecular species in developing bovine enamel. J Biol Chem 255:9760–9768
2. Robinson C, Kirkham J (1985) Dynamics of amelogenesis as revealed by protein compositional studies. In: Butler WT (ed) The chemistry and biology of mineralized tissues. Ebsco Media, Birmingham, pp 248–263
3. Smith CE, Pompura JR, Borenstein S, Fazel A, Nanci A (1989) Degradation and loss of matrix proteins from developing enamel. Anat Rec 224:292–316
4. Fincham AG, Belcourt AB, Termine JD (1982) Changing patterns of enamel matrix proteins in the developing bovine tooth. Caries Res 16:64–71
5. Suga S (1970) Histochemical observations of proteolytic enzyme activity in the developing dental hard tissues from the rat. Arch Oral Biol 15:555–558
6. Shimizu M, Tanabe T, Fukae M (1979) Proteolytic enzyme in porcine immature enamel. J Dent Res 58:782–789

7. Moe D, Birkedal-Hansen H (1979) Proteolytic activity in developing bovine enamel. J Dent Res 58:1012–1013
8. Crenshaw MA, Bawden JW (1984) Proteolytic activity in embryonic bovine secretory enamel. In: Fearnhead RW, Suga S (eds) Tooth enamel IV. Elsevier, Amsterdam, pp 109–113
9. Carter J, Smillie AC, Sheperd MG (1984) Proteolytic enzyme in developing porcine enamel. In: Fearnhead RW, Suga S (eds) Tooth Enamel IV. Elsevier, Amsterdam, pp 229–233
10. Carter J, Smillie AC, Sheperd MG (1989) Purification and properties of a protease from developing porcine enamel. Arch Oral Biol 34:195–202
11. Overall CM, Limeback H (1988) Identification and characterization of enamel proteinases isolated from developing enamel. Biochem J 256:965–972
12. DenBesten PK, Heffernan LM (1989) Separation by polyacrylamide gel electrophoresis of multiple proteases in rat and bovine enamel. Arch Oral Biol 34:399–404
13. Aoba T, Fukae M, Tanabe T, Shimizu M, Moreno EC (1987) Selective adsorption of porcine amelogenins onto hydroxyapatite and their inhibitory activity on hydroxyapatite growth in supersaturated solutions. Calcif Tissue Int 41:281–289
14. Silness J, Gustavsen F (1970) Some observations on the fine structure of fluorosed enamel. Acad Odontol Scand 28:701–720
15. Shinoda H (1983) Effects of long-term administration of fluoride on enamel formation in rats. In: Suga S (ed) Mechanisms of tooth enamel formation. Quintessence, Tokyo, pp 273–284
16. DenBesten PK, Crenshaw MA (1984) The effects of chronic high fluoride levels on forming enamel in the rat. Arch Oral Biol 29:675–679
17. DenBesten PK (1986) Effects of fluoride on protein secretion and removal during enamel development in the rat. J Dent Res 65:1272–1277
18. Trautschold I, Werle E, Zickgraf-Rüdel G (1967) Trasylol. Biochem Pharmacol 16:59–72
19. Scocca J, Lee YC (1969) The composition and structure of the carbohydrate of pineapple stem bromelain. J Biol Chem 244:4852–4863
20. Bobb D (1972) Isolation of stem bromelain and its partial characterization by gel electrophoresis. Prep Biochem 2:347–354
21. James GT (1978) Inactivation of the protease inhibitor phenylmethylsulphonyl fluoride in buffers. Anal Biochem 86:574–579
22. Sorensen K, Brodbeck U (1986) A sensitive protein assay method using microtiter plates. Experientia 42:161–162
23. Church FC, Porter DH, Gatignani GL, Swaiswood, HE (1985) An *o*-phthalaldehyde spectrophotometric assay for proteinases. Anal Biochem 146:343–348
24. Laemmli UK (1970) Cleavage of structural proteins during the assembly of the head of bacteriophage T4. Nature 227:680–685
25. Snedecor W (1959) Statistical methods. Iowa State College Press, Ames

Enamel Formation During Porcine Amelogenesis

Takaaki Aoba[1]

Summary. The present work was aimed at elucidating the mechanism of enamel mineralization during porcine amelogenesis by specifically investigating (1) the driving force for precipitation in the fluid microenvironment from which enamel crystals form, (2) changes in the property of matrix proteins, particularly the amelogenins, originating from their post-secretory degradation, and (3) regulatory mechanisms of crystal growth related to the interaction between these proteins and enamel mineral. Analyses of ionic concentrations (and activities) of the fluid separated from secretory porcine enamel indicated that enamel mineralization during the secretory stage of porcine amelogenesis occurs in a specific microenvironment segregated from the circulating blood. It was also suggested that the driving force for precipitation at this stage is controlled by the cellular transport of lattice ions (particularly Ca ions) and the presence of Ca-binding ligands (regulating calcium ion activity) in the fluid. Our in vitro work also showed that the originally secreted porcine amelogenin (25 kd by SDS-PAGE), as well as 60–90 and 32 kd non-amelogenins, can adsorb selectively onto apatitic surfaces. Importantly, this adsorption of amelogenin is intimately related to the inhibitory activity of crystal growth of hydroxyapatite in supersaturated solutions and regulation of isotopic exchange of ions (e.g., calcium) on the crystal surfaces in saturated conditions. This potential function of amelogenins in situ seems to be modulated by enzymatic cleavages of specific segments at the N- and C-termini.

Key words: Biomineralization—Amelogenesis—Enamel Mineral—Crystal Growth—Amelogenins

Introduction

Enamel mineralization during amelogenesis is generally divided into the secretory, transitional, and maturational stages on the basis of changes in mineral and protein contents and cellular structures [1–3]. Considerable work using various techniques and animal models has been conducted on the

[1] Forsyth Dental Center, 140 Fenway, Boston, MA 02115, USA

individual stages of amelogenesis. The present consensus is that mineralization during amelogenesis is controlled by cellular activities (transport of mineral lattice ions and secretion and removal of organic matter). Autoradiographic studies have shown that ionic components in a circulating blood, e.g., Ca ions, can be transported rapidly into the forming enamel tissue [4]. Bawden et al. [5] demonstrated that the calcium transport in the secretory stage was regulated by the ameloblast, in contrast to the transport of phosphate, while no regulation of transport of either ion was observed at the maturational stage. However, there is a paucity of information about the microenvironment surrounding the precipitating enamel crystals and the driving force for precipitation during amelogenesis. Also, little is known about the nature, stoichiometry, and thermodynamic property of enamel crystals formed at various developmental stages.

The enamel matrix proteins, secreted by the ameloblasts during early stages of amelogenesis, are generally categorized into 2 groups, amelogenins and enamelins (or non-amelogenins), based on their amino acid compositions, solubility properties, and apparent molecular weights [6–8]. Studies using molecular biological approaches also suggested that the amelogenins and the enamelins are distinct classes of proteins [9]. The group of enamelins is minor matrix constituents and, recently, questions have been raised about the nature and origin of this group of proteins [10]. In contrast, the amelogenins are major matrix constituents and the primary structure of the secreted amelogenins has already been determined in various mammalian species [9, 11–13].

The most distinctive feature of amelogenesis of mammals, in contrast to their dentinogenesis and osteogenesis, is that the secreted matrix proteins are removed, almost completely, during the developmental stages, leaving large apatite crystallites in a well oriented fashion. It is most likely that the remarkable length and orientation of enamel crystallites are the results of the ameloblast-mediated matrix-mineral interaction. Recently, Weiner [14] postulated, on the basis of features in tooth, bone, and mollusc shell formation, that 2 kinds of proteins, hydrophilic (acidic in nature, aspartic acid-rich and, in some cases, phosphoserine-rich) and hydrophobic (proline or hydroxyproline-rich), are always involved in the process of "organic matrix-mediated" biomineralization. As to the function of these molecules, it has been proposed that the acidic proteins, bound to the hydrophobic proteins (or rigid substrates in vitro), play a role as nucleators but can also act as crystal growth inhibitors when free in solution [15]. Analogous roles have also been advanced for enamel proteins. In the literature, it is widely accepted that the enamelins are intimately associated with enamel apatite, while the amelogenins occupy the space between the crystals [3]. It is also proposed that enamelin-like proteins are highly conserved during vertebrate evolution [16] and that these anionic macromolecules may serve a primary function in the initiation of calcium apatite formation [17]. However, the initial precipitation during mammalian amelogenesis is induced following dentin mineralization. Furthermore, there is no direct evidence for the binding of enamelins to some of the hydrophobic proteins (amelogenins) or substrates, thus playing a role as nucleators during the secretory stage of amelogenesis. On the basis of the foregoing consideration, it is most likely that the secreted matrix proteins, including amelogenins and enamelins, function in a different way in enamel mineralization than do matrix proteins in dentinogenesis or osteogenesis.

Our working hypothesis about the mechanism of enamel mineralization at early stages of amelogenesis is that: (1) the nature and precipitation kinetics of enamel mineral are primarily determined by the degree of saturation of the fluid phase, which is controlled by the supply of lattice ions (especially calcium transport) through the ameloblast layer, (2) with the intermittent supply of calcium ions from the cell layer, the driving force in the fluid phase becomes high enough to induce rapid precipitation of calcium phosphate salts in the vicinity of the ameloblast, (3) the secretion of amelogenins and enamelins prevent further precipitation through their adsorption onto the formed crystal surfaces, (4) their functions as inhibitors of crystal growth are modulated through the degradation (i.e., cleavages of specific segments) of these macromolecules prior to their massive removal, and (5) the Ca-ligands complex in the fluid acts as a reservoir of the ion (through dissociation), maintaining a modest degree of supersaturation for a slow growth of the formed apatite crystals in the underlying secretory enamel, concomitant with the degradation of the inhibitory proteins. The present work was planned to gain experimental evidence for this overall hypothesis.

Materials and Methods

Preparation of Enamel Samples

The enamel used in this work was obtained from freshly slaughtered piglets. The main reason for the selection of this animal model is that the permanent teeth yield sufficient amounts of enamel tissue at various stages of amelogenesis. The incisors and molars (with the attached enamel organ tissues to prevent contact with blood) were dissected out of the mandible of 5- to 6-month-old animals. The incisors were used to dissect the secretory enamel layer-by-layer from the outermost (young) toward the innermost (older) zones, as reported previously [18]. Prior to the enamel dissection, the covering soft tissues were removed. The exposed enamel surface was wiped gently with a paper tissue to remove the cellular debris and traces of blood. The outermost enamel, about $100\,\mu m$ in thickness, was scraped carefully from the labile surface of the secretory tissue with the aid of a razor blade. Following this separation, the underlying secretory enamel tissues were separated into four layers (each $200-300\,\mu m$ in thickness) from the outer toward the dentin-enamel junction (DEJ). The enamel samples thus dissected into 5 layers were individually weighed and then freeze-dried. The molars were used for dissection of the secretory enamel and then in the separation of the fluid.

Separation of Enamel Fluid

The fluid separation was commenced immediately after the dissection of the enamel organs. The covering soft tissues were removed as soon as possible to minimize the flux of ions into (or out of) the enamel tissue through the cellular membrane. The exposed surface of the secretory enamel was covered with mineral oil. The mineral oil used was presaturated with water vapor by bubbling 1.8/98.2 CO_2/N_2 gas that had passed through a 160 mM NaCl solution. The secretory enamel was dissected with the aid of an excavator and pooled under the mineral oil in a 5 mL-volume ultracentrifuge tube (Polyallomer, Calif.). Separation of the fluid from the solid tissue was carried out by ultracentrifugation at 2.4×10^5 g. The yield of the fluid was about 45 to 55 μL from 1 g of secretory enamel. Pooled fluid samples were used for measurements of the concentrations of total Ca, Mg, Na, K, total P, HCO_3 (ion chromatography, Dionex 4010i with a conductivity detector), Cl, and F (in conjunction with ion-selective electrodes). The pH of the fluid was measured in an oil bath at room temperature (about 25°C), using needle pH and reference electrodes (Microelectrodes Inc. N.H.), as reported previously [19]. The calcium activity of the fluid was measured with a specific Ca ion microelectrode (Ionetics Inc., Calif.). In order to assess the ionic strength of the fluid, the osmolality was determined in a Nanoliter Osmometer (Clifton, N.Y.).

Preparation of Enamel Matrix Proteins

The preparation of enamel proteins was conducted in 3 different ways for specific purposes. (1) To collect whole matrix proteins from secretory porcine enamel, the dissected tissues were dissolved in 0.5 M acetic acid containing protease and phosphatase inhibitors [20]. The proteins extracted in the acid solution were desalted and concentrated by ultrafiltration using a YM-5 membrane (Amicon; M.W. cutoff 5 kd). (2) To separate the matrix proteins based on their hydrophobic/hydrophilic properties, the extraction was conducted using chloroform/methanol 3:1. The dissected secretory enamel was added to a centrifuge tube containing the organic solvent (solid/solution = 1/40 wt/v) and then well dispersed with the aid of a sonic dismembrator. The slurry was separated by centrifugation. The supernatant was removed carefully; the proteins extracted in the solvent were concentrated and then freeze-dried. The sedimented solid was dissolved in 0.5 M acetic acid and the proteins dissolved in the acid solution were recovered as described above. (3) To purify the amelogenins having various molecular weights, the matrix proteins extracted according to (1) were further fractionated using gel filtration chromatography through a Sephadex G-100 column equilibrated with 50 mM carbonate buffer (pH 10.8) in the presence of protease- and phosphatase-inhibitors. Each protein fraction was concentrated by ultrafiltration. Further purification of the amelogenins was carried out by recycling the chromatography 5 times under the same conditions or using ion-exchange columns [21]. Following the final chromatography, eluent fractions corresponding to the desired proteins were pooled, desalted using the ultrafiltration membranes, and then lyophilized. The purity of the separated protein samples was verified by SDS-PAGE and amino acid analysis, and by determining their N- and C-terminal sequences.

Adsorption Experiments

The adsorbent used was hydroxyapatite (6.4 m^2/g) which was prepared and characterized in our laboratories. The matrix proteins separated from the whole secretory enamel were used as adsorbates. Adsorbate-adsorbent equilibration was conducted in acetate buffer (pH 6.0) at 0°C, as reported previously [20]. The adsorbates were solubilized in the buffer to yield a 0.1% wt/v protein solution. Accurately weighed amounts of the adsorbent, 5–100 mg, were added to 2 mL of the adsorbate solution. The slurry was mixed continuously in a rotator for 1 h at the specified temperature and then separated by centrifugation at 10,000 g for 10 min. The supernatant was recovered carefully and then another 2 mL of the same solvent used for equilibration was added to the centrifuge tube containing the pellet. The crystals were dispersed with a vortex mixer, centrifuged again, and then the supernatant was removed. This washing procedure was repeated 3 times to minimize the contamination of unadsorbed macromolecules or to remove species non-specifically adsorbed onto the crystals. Following the final washing, the sedimented crystals were completely dissolved in 0.5 M acetic acid. All procedures for protein recovery were conducted at 0°C unless otherwise specified. The acid solution containing the adsorbed proteins was desalted by passing it through a PD-10 column (Pharmacia-LKB, N.J.), using 0.5 M acetic acid as an eluent, and then freeze-dried. The unadsorbed proteins remaining in the equilibrating solvent and the proteins in washing solutions were also desalted in the same way and then freeze-dried. The amounts of proteins adsorbed were determined by dissolving the freeze-dried samples in 20 mM acetic acid and analyzing the total protein concentration by Lowry's method or amino acid analysis. The protein moieties recovered from the crystal surface and the proteins remaining in the equilibrium solution (i.e., the first supernatant) were characterized by SDS-PAGE using 15% to 18% polyacrylamide gels.

Effect of Protein-Crystal Interaction on Ca Incorporation on the Crystals

In this work, our interest was focused on the coatings of apatite crystals with enamel proteins in dispersed forms. Since the amelogenins were sparingly soluble under physiological conditions but became soluble in low ionic strength solutions at cool temperatures [8], the whole matrix proteins extracted from the secretory enamel (mainly the amelogenins) were solubilized in deionized water at 0°C to yield 0.2% wt/v adsorbate solution. The protein solution was mixed with the same volume of 0.32 M NaCl solution; the resulting solution (0.1% wt/v proteins) had an ionic strength of 160 mM. In this solution and at room temperature, the proteins commenced to aggregate, resulting in a clouding of the solution. Hydroxyapatite crystals (the same batch as used in the adsorption experiments), in various amounts to provide total surface areas covering the range of 0.01 to 0.2 m^2, were introduced into the protein solution and well mixed with a vortex. The protein-crystal equilibration was continued for 1 h at room temperature. Following the equilibration, the slurry was separated by centrifugation, as described above. The sedimented (protein-coated) crystals were used to investigate the incorporation of calcium ions onto the protein-coated crystals in a well-defined solution. The experimental solution (100 mL in volume) had an ionic composition similar to that of the enamel fluid (pH 7.3, total phosphate = 3.9 mM, total CO_3 = 10 mM, and background electrolyte = 150 mM as NaCl) but with a lower total calcium concentration (0.10 mM) to yield about the same calcium activity of the fluid and saturation with respect to hydroxyapatite. The solution was maintained at 37° ± 0.2°C in a water bath and well stirred with a magnetic stirrer. Trace amounts of $^{45}CaCl_2$ (10 μCi) were added to the solution. The prepared protein-coated crystals were dispersed in 1 mL of 160 mM NaCl and then introduced into the experimental solution (time zero). Time courses of the changes in radioactivity in solution were monitored by taking aliquots (1 mL) before the addition of the crystals and at timed intervals over the experimental period (5 h). The withdrawn aliquot was transferred to a microcentrifuge tube and then centrifuged at 10,000 g for 1 min. The radioactivity in the supernatant was counted using a liquid scintillation counter. The results (kinetics of incorporation and the incorporated amount at equilibrium) were compared between the experiments using the protein-free crystals and the protein-coated crystals. The experimental solution used was close to saturation with respect to the crystals. Nevertheless, in all experiments, it was verified that no appreciable changes in the total Ca concentration of the experimental solution occurred before and after equilibration. In parallel experiments, but without the addition of the

Table 1. Ionic composition and degree of saturation of enamel fluid

Analytical values		*Calculated values*		*Degree of saturation*[a]
Total Ca	0.50 mM	Free Ca^{2+}	1.53 mM	$(IP/K_{HA})^{1/9}$
Ca^{2+} Activity	0.053 mM	$CaHPO_4^0$	1.18×10^{-2} mM	= 3.56
Total P	3.90 mM	$CaH_2PO_4^+$	2.38×10^{-4} mM	
Total Mg	0.83 mM	$CaHCO_3^+$	8.33×10^{-3} mM	$(IP/K_{OCP})^{1/8}$
Total HCO_3	10 mM	$CaCO_3^0$	5.97×10^{-4} mM	= 0.59
Na + K	160 mM	Ca-lactate^0	1.71×10^{-2} mM	
Cl	130 mM	Bound Ca^0	0.31 mM	$(IP/K_{EN})^{1/7.8}$[b]
Lactate	11 mM	Free HCO_3^-	8.64 mM	=0.93 to 1.21
pH	7.3	Pco_2	0.0184 atm	
Osmolality[c]	312	Ionic strength	165 mM	

[a] IP stands for the ionic activity product of the salt concerned and the degree of saturation is expressed in terms of the mean ionic activity products

[b] The adopted stoichiometries of the outer and inner secretory porcine enamel were $[Ca]_{4.16}[HPO_4]_{0.41}[CO_3]_{0.42}[PO_4]_{2.16}[OH]_{0.16}$ and $[Ca]_{4.29}[HPO_4]_{0.31}[CO_3]_{0.40}[PO_4]_{2.29}[OH]_{0.29}$, respectively

[c] mosmol/kg H_2O

radioactivity, the protein-coated crystals were incubated for 5 h in the experimental solution and, at the end, the slurry was separated by centrifugation. The proteins remaining on the crystal surfaces and the proteins dissolved in the experimental solution were individually recovered, according to the same procedures as used in the adsorption studies, and then characterized using SDS-PAGE.

Electron Microscopy

Part of the secretory enamel tissue was lyophilized immediately after dissection and then embedded with POLY/BED 812. Ultra-thin sections were prepared with a diamond knife and the morphology of the crystals, without staining, was observed in either a JEM 1200EX electron microscope at 100 kV or a JEM-2010 electron microscope at 200 kV.

Results and Discussion

Table 1 shows the results of the chemical composition of the separated fluid. A significant finding was that the ionic calcium activity, $(Ca^{2+}) = 0.053$ mM, was one order of magnitude lower than the total calcium concentration, suggesting the presence of ligands complexing the calcium ions. Additionally, the fluid contained about 10 mM total HCO_3, which was much higher than the total concentrations of lattice ions, $[Ca] = 0.5$ mM and total $[PO_4] = 3.9$ mM. At the determined pH value of the fluid, it was found that the partial pressure of CO_2 of the fluid corresponded to 0.018 to 0.02 atm which is less than that (0.05 atm) of porcine serum. The foregoing results indicate that enamel mineralization during the secretory stage of amelogenesis occurs under a distinct, compartmentalized microenvironment, isolated from the circulating blood (at least in relation to some of the common constituents), although the fluid was kept isosmotic with serum, thus preventing undesirable water flows across tissues. Previously, we reported the estimated degree of saturation (DS) of the fluid with respect to calcium phosphate salts of interest, which were calculated using analytical data available at that moment [19]. Since the concentration of bicarbonate and organic acids (e.g., the total concentration of lactate of the fluid was 11 mM [22]) were analytically determined after that publication, the DS of the fluid was recalculated by taking into consideration the pertinent ion pairs with these species. A new mass balance for calcium considered was:

$$\text{Total Ca} = \text{free } Ca^{2+} + CaHPO_4^0 + CaH_2PO_4^+ + CaHCO_3^+ + CaCO_3^0 + \text{Ca-Lactate}^+ + \text{bound } Ca^0$$

Since the measurement of pH of the fluid was conducted at 25°C, the numerical values of the solubility product for hydroxyapatite (K_{HA}) and octacalcium phosphate (K_{OCP}) are 2.63×10^{-60} and 1.05×10^{-47}, respectively (detailed in [19]). Recently, we reported the solubility product constants for secretory porcine enamel (K_{EN}), which were determined under controlled Pco_2 = 0.018 atm (N_2-CO_2 mixtures). The values of K_{EN} obtained for the outer and inner secretory enamel on the basis of the stoichiometry model $[Ca]_{5-x}[HPO_4]_v[CO_3]_w[PO_4]_{3-x}[OH]_{1-x}$, were 1.45×10^{-40} and 1.08×10^{-43}, respectively [23]. In Table 1 are also shown the calculated ionic composition of the fluid and the DS values with respect to HA, OCP, and the enamel minerals. The results indicate that the calcium-ligand complex accounts for about 62% of the total calcium, significantly affecting the DS of the fluid. The DS of the fluid separated from the secretory enamel is essentially in equilibrium with the forming enamel apatites, and slightly supersaturated and undersaturated with respect to HA and OCP, respectively.

Concerning the Ca-ligands, our previous studies [24] showed that the apparent dissociation constant of the assumed complex was about 7.5×10^{-5} mol/L. This value indicates that the strength of the Ca-ligand binding is rather modest; the determined value was significantly greater than the dissociation constant (1.95×10^{-6} mol/L) of Ca-binding by the whole organic matrix of fetal bovine enamel, reported by Drinkard et al. [25]. This feature seems advantageous if complexed calcium acts as a reservoir (a fine regulatory system through association-dissociation) of calcium levels in the fluid during the secretory stage of amelogenesis. Our computer simulation of changes of fluid composition taking into consideration the presence of Ca-ligands, clearly showed that free Ca^{2+} concentrations in fluid can increase from 0.15 mM to 0.2 mM with an increase in total calcium from 0.5 mM to 1 mM (Fig. 1.A); this increase in free Ca^{2+} concentration is enough to induce supersaturated conditions with respect to OCP (Fig. 1.B).

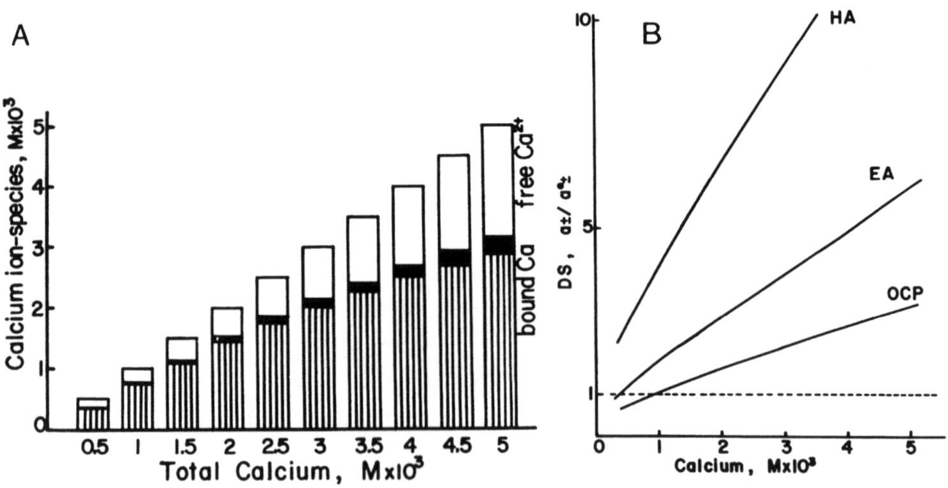

Fig. 1. **A** Results of calculation of calcium ionic species in fluid and **B** the degree of saturation (*DS*) of fluid with respect to hydroxyapatite (*HA*), octacalcium phosphate (*OCP*), and enamel mineral at the secretory stage (*EN*) as a function of the increment of the total calcium concentration of the fluid

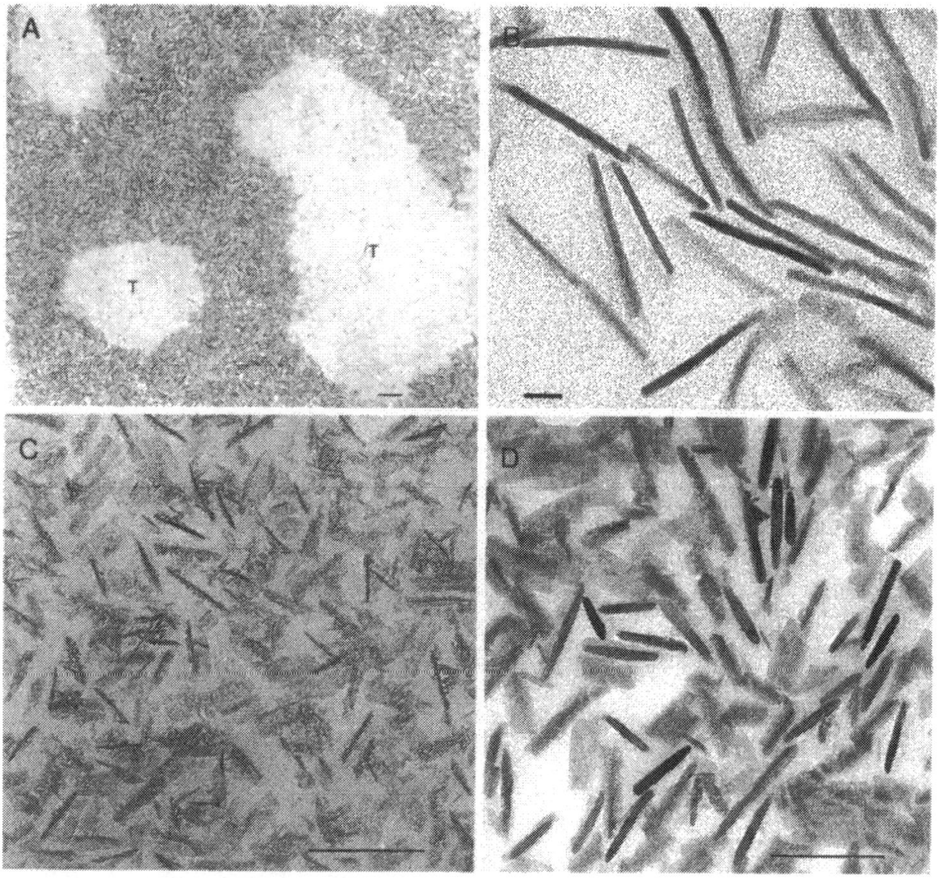

Fig. 2. Electron micrograms of porcine enamel crystals. **A** Thin ribbons precipitated around Tomes processes (*T*) of the ameloblast. **B** High resolution images of these crystals having thicknesses corresponding to several unit cells of apatite crystals. **C** The crystals found in the outer region, displaying non-uniform deposition of aptitic unit cells on the lateral planes of thin ribbons. **D** The grown crystals found in the middle region of secretory enamel; note the presence of a *central dark line. Thin and thick horizontal bars* represent 100 nm and 10 nm, respectively

Microradiographic studies by Suga [1] clearly show that mineralization within the secretory stage of mammalian amelogenesis progresses from the outer toward the inner regions. This advancement of mineralization is due to an increase in the size of the crystallites (i.e., crystal growth) (Fig. 2). The crystals precipitated in the vicinity of the ameloblast were thin ribbons in appearance (Fig. 2.A,B) and their initial growth process was characterized by a non-uniform accretion of apatitic unit cells on the lateral planes of a thin ribbon (Fig. 2.C). The grown crystallites, found in underlying zones, retained defects in their core region, the so-called central dark-line images (Fig. 2.D). Several investigators [26] have advocated that these features of enamel crys-

```
       1         45 46  62 63                    148    161           173
25 kd  M----------W-L---H-A-------------------S-----W-PATDKTKREEVD

23 kd  M-------------------------------------------------W

20 kd  M-----------------------------------------S

       M---- 7 kd ------H A------ 11 kd -----S

       M-5 kd -W L---------- 13 kd ---------S
```

Fig. 3. The 25 kd porcine amelogenin secreted by the ameloblast and its degraded products

tallites (i.e., thin ribbons and central defects) may be ascribed to possible involvement of OCP-like phases as precursors. Interestingly, the observation of initial precipitates in in vitro carbonatoapatite preparations [27] disclosed that all the features (i.e., thin-ribboned in appearance, the central dark lines, and the non-uniform deposition) were reproduced in the neutral solutions, whereas rod-like apatite crystals, having hexagonal cross sections and without any vestiges of the central dark lines, were formed in the presence of high concentrations of fluoride or at pH 10, where the precipitation of OCP becomes kinetically unfavorable. However, no crystallographic evidence for OCP precipitation in mammalian amelogenesis is reported in the literature.

A central issue in enamel mineralization during amelogenesis is to understand how the observed precipitation and subsequent growth of enamel crystals are regulated through the protein-mineral interaction. Our interest was focused on the role played by the amelogenins. The whole sequence of the intact porcine amelogenin (25 kd by SDS-PAGE) was reported by Yamakoshi et al. [13]. This amelogenin protein, comprising 173 amino acid residues, has a unique sequence, i.e., all the charged residues are located within the N-terminal 40 and C-terminal 15 residues; particularly, the highly charged segment composed of 2 Glu, 2 Asp, 2 Lys, and 1 Arg is found at the C-terminus. It is also established that this intact protein is degraded specifically by proteolytic enzymes shortly after secretion [7], giving rise to a complex picture of protein constituents in the forming tissue, such as the moieties of 23, 20, 13, 11, 7, and 5 kd. In Fig. 3 are illustrated the sites where the enzymatic cleavages of the 25 kd porcine amelogenin takes place, according to the work by Shimizu and Fukae [8] at the Tsurumi University, Yokohama, Japan.

Figure 4 shows SDS-electrophoretograms of the proteins separated from the outermost (close to the ameloblast) through innermost (close to DEJ) zones of secretory porcine enamel. A comparison of the SDS-PAGE patterns indicated that appreciable changes in the protein composition take place within the secretory stage. The 25 kd amelogenin, as well as high molecular weight proteins (60–90 kd), were concentrated in the outermost secretory enamel. Most of the 60–90 kd proteins were assigned as being non-amelogenins [28]. In the underlying inner zones, the 20 kd amelogenin increased in relative terms as

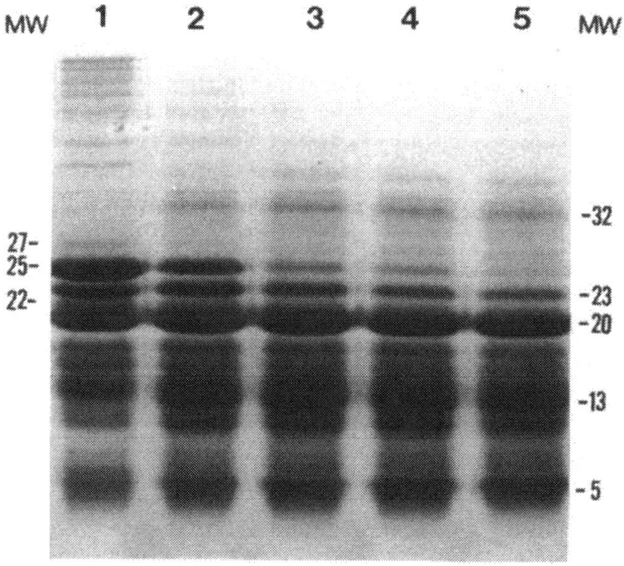

Fig. 4. SDS-electrophoretograms of enamel matrix proteins extracted from the various layers (I through V) of secretory porcine enamel. Note the presence of 27, 25, and 22 kd amelogenins, as well as the 60–90 kd non-amelogenins, in the outermost zone (close to the ameloblast). With substantial decreases of these moieties, the 20 kd and lower molecular weight amelogenins, as well as the 32 kd non-amelogenins, increased in relative terms in the underlying zones.

the 25 kd amelogenin decreased. The 11–13 and 5–7 kd amelogenin fragments also increased in the inner zones. In addition, as the 60–90 kd non-amelogenins decreased substantially, the 32 kd band became prominent. This 32 kd group (at least 7 constituents having pl = 3–4.5 were distinguished) has a unique N-terminal sequence, LxHVPGRIPPGYGRPPT-, which did not correspond to the sequence of porcine amelogenin [29]. It is notable that two faint bands at 27 and 22 kd were discerned from the proteins separated from the outermost secretory enamel and that the corresponding proteins were not discernible in the underlying tissue. In studies using antibodies raised against a synthetic peptide corresponding to the C-terminal 12 residue segment of the 25 kd porcine amelogenin, it was found that these 27 and 22 kd moieties reacted with the antibodies as the 25 kd protein did, indicating the presence of the common epitope at the C-terminus [30]. This finding supports the view that the ameloblast secrete amelogenin families through splicings of DNA information [9, 11].

The foregoing results on protein localization in secretory porcine enamel indicate that the 25 kd intact amelogenins (and the 27 and 22 kd proteins) are unstable in situ. Our recent studies using ^{1}H-NMR and photo-CIDNP (chemically induced dynamic nuclear polarization) [31] showed that the C-terminus of the 25 kd porcine amelogenin is exposed on the molecular

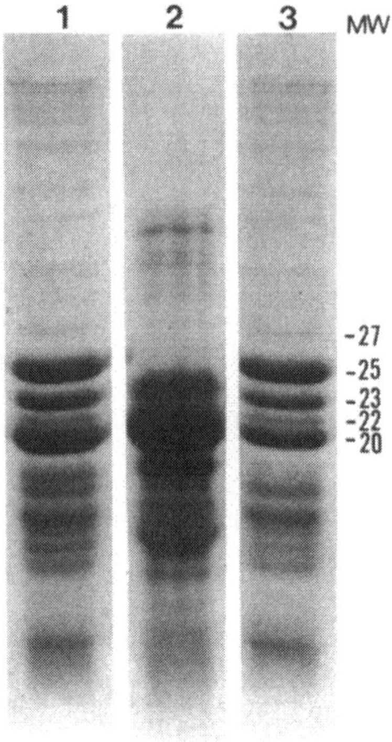

Fig. 5. SDS-electrophoretograms showing the extraction of enamel matrix proteins with chloroform/methanol 3:1 (v/v). *Lane 1*, the whole matrix proteins isolated from the outer secretory enamel; *lane 2*, proteins extracted in the organic solvent; *lane 3*, proteins recovered on the crystals after extraction. Note the remarkable difference in the hydrophobic/hydrophilic nature between the 25 kd intact amelogenin (and 27 and 22 kd proteins) and the degraded products (23 and 20 kd) lacking the hydrophilic segment at the C-terminus

surface, supporting the contention that the segment at the C-terminus is easily accessible to enzymatic cleavage. An intriguing question is whether the presence of the hydrophilic segment at the C-terminus and its cleavage are essential parts of the regulatory mechanism of enamel mineralization at early developmental stages. In the present studies, we first tested a possible influence of the C-terminal segment on the hydrophilic/hydrophobic nature of the amelogenin molecules by investigating the solubility of the proteins in the organic solvent (chloroform-methanol 3:1). The pertinent SDS-PAGE patterns are shown in Fig. 5. The most prominent findings were that (1) the 25 kd amelogenin, as well as the high molecular weight non-amelogenins, was sparingly soluble in this solvent, (2) the 27 and 22 kd moieties having a common epitope at the C-terminus were also sparingly soluble, and (3), in contrast, the 23 and 20 kd amelogenins, lacking the hydrophilic segment at the C-terminus, were extracted preferentially in the solvent. These results provide evidence that the property of the amelogenins is actually modified by the cleavage of the C-terminal segment.

As cited in the Introduction, it is widely accepted that the enamelins bind tightly onto the enamel mineral, while the amelogenins are segregated from the crystal surface during mineralization [3]. However, based on the published primary structures of the parent amelogenins showing the presence of a hydrophilic sequence at the C-terminus (and the observed difference in

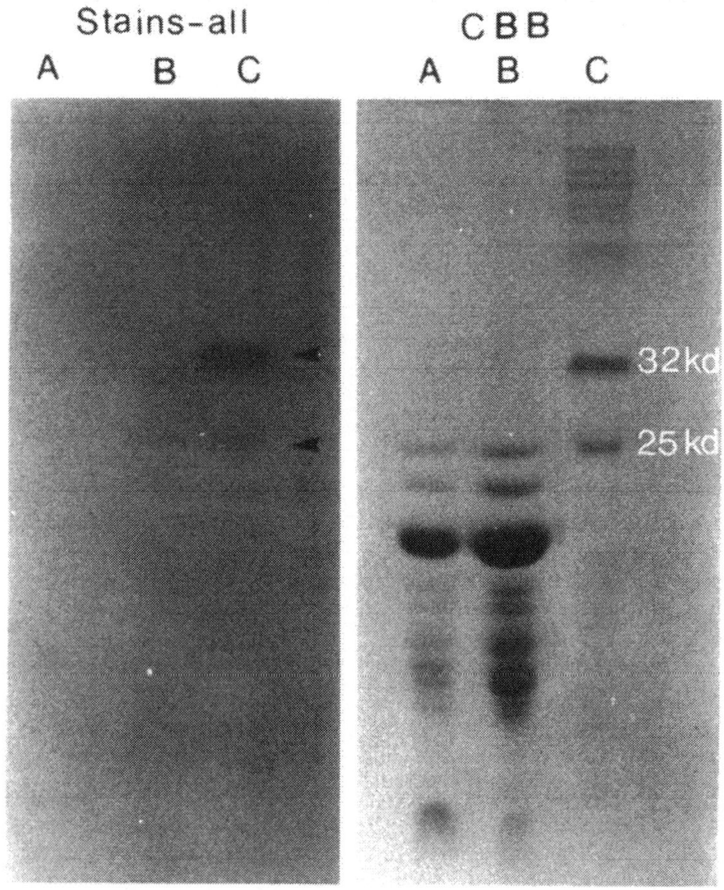

Fig. 6. SDS-electrophoretograms showing the selective adsorption of porcine enamel proteins onto hydroxyapatite. *Lanes A*, the original enamel proteins used as adsorbates, which were isolated from the secretory enamel; *lanes B*, proteins remaining in solution after equilibration with the adsorbent crystals; *lanes C*, proteins adsorbed onto the crystals. The left- and right-side lanes A through C were stained with Stains-all and Coomassie brilliant blue (*CBB*), respectively. Note the staining (blue in color) of the 32 and 25 kd proteins with Stains-all

hydrophilic/hydrophobic property between the intact amelogenin[s] and its degraded products), we hypothesized that the intact amelogenins may play a different role in the process of enamel mineralization. This hypothesis was tested in in vitro adsorption studies. Figure 6 shows typical results of the selective adsorption of enamel proteins onto hydroxyapatite. The original adsorbate shown in lane A was the whole matrix proteins separated from secretory porcine enamel. As shown in lanes B and C, whereas the degraded products of the 25 kd amelogenin (e.g., 23 and 20 kd proteins) were not adsorbed appreciably onto the crystals and most of them remained in the solution

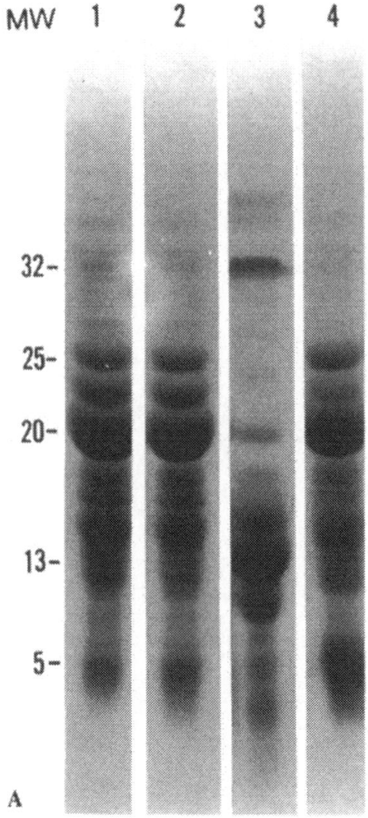

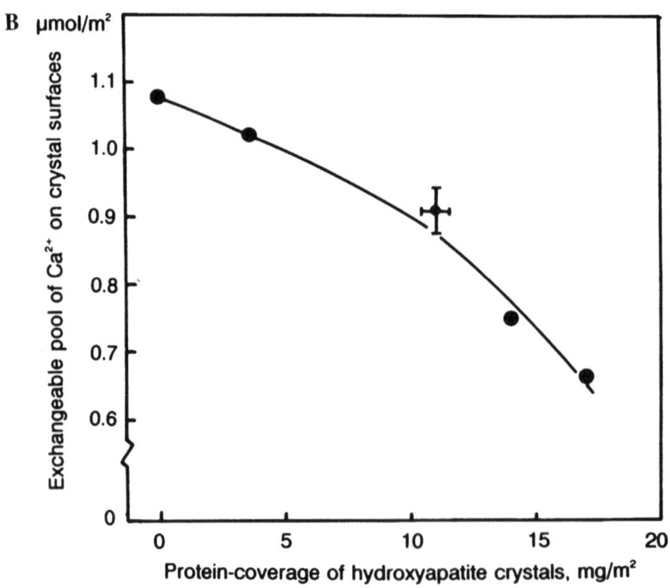

Fig. 7. A SDS-electrophoretograms showing coating hydroxyapatite with enamel matrix proteins in dispersed forms. *Lane 1*, proteins used as adsorbates; *lane 2*, proteins sedimented with the crystals in 160 mM NaCl solution at room temperature; *lane 3*, proteins dissolved in the fluid-simulated solution during the isotopic exchange reaction with calcium ion; *lane 4*, proteins remaining on the crystal surfaces at the end of the reaction. **B** Changes of the exchangeable Ca pool on the crystal surface as a function of the protein-coating of the crystals

(lane B), the 25 kd intact amelogenin, as well as the 32 kd and 60–90 kd non-amelogenins, were concentrated on the crystals after equilibration (lane C). The left-side lanes of Fig. 6 were stained with Stains-all. Interestingly, the band corresponding to 32 kd proteins was stained most extensively (blue in color), while the 25 kd amelogenin was also stained faintly. The obtained staining of the 32 kd proteins is consistent with the finding that this protein group was highly phosphorylated (about 7 mols P per mol protein) [29]. Similarly, the present results suggest that some of the 25 kd proteins may be phospholyrated. In this connection, previous studies using two-dimensional electrophoresis [32] demonstrated that the 25 kd porcine amelogenin is separated into charge families, one major moiety at about pI = 6.2 and at least three or four variants having different pI values between 6 and 6.5. However, no quantitative data about the phospholyration of porcine amelogenins (and evidence for the phospholyration site) are available at present.

An important finding relevant to possible functions of the amelogenins in situ [18, 20] is that the observed adsorption of the 25 kd amelogenin onto hydroxyapatite is related intimately to the inhibition of precipitation kinetics of the crystals in supersaturated solutions. Furthermore, the present studies provided an insight into possible functions of enamel matrix proteins (in aggregated forms) in solution having a composition similar to that found in the enamel fluid. Figure 7 shows the effect of the protein-coating on the Ca incorporation onto the crystals. Following the proteins-crystal equilibration in solution having an ionic strength of 160 mM at room temperature, most of the proteins were sedimented with the crystals (lane 2). When the protein-coated crystals were added to the fluid-simulated solution for investigating Ca exchange at the crystal-solution interface, it was found that appreciable amounts of the proteins (approximately 20% of the total proteins originally coated on the crystals) were dissolved in the solution during the experimental period. The moieties which dissolved corresponded to the 32 kd non-amelogenin and the 20 and 13–11 kd amelogenins (lane 3), but the 25 and 23 kd amelogenins were barely dissolved. As illustrated in Fig. 7.B, the exchangeable pool of calcium on the crystal surface decreased with an increase of the protein coating per unit surface area of the crystals. These results, as well as previous findings with respect to fluoride incorporation into apatite lattice [33], indicate that coating hydroxyapatite crystals with enamel matrix proteins brings about a retardation of ion incorporation into the crystals. It is pertinent to point out that the expected ratio of protein/mineral in the secretory enamel is substantially higher than those ratios tested in the present study. Therefore, very significant reductions in the effective surface areas of enamel crystals are anticipated in situ, thereby retarding ion incorporation into the crystalline lattice and, by exchange, on the crystal surface.

In relation to the mechanism of the amelogenin-crystal interaction shown above, it is a logical proposition that the obtained properties (i.e., adsorption affinity and inhibitory activity) of the intact amelogenins are related to their molecular structures (i.e., presence of the functional sequences and radicals, such as phosphorylation, and the molecular conformations). In this connection, it is a significant feature of mammalian amelogenins that their primary structures display a high degree of homology, particularly the first 30 and last 10 residues at the N- and C-termini [7, 11]. The observation that the 23 kd amelogenin, lacking the C-terminal 12 residues, already substantially decreases the adsorption affinity onto the crystals (Fig. 6), suggests that the cleaved segment at the C-terminus may be the functional sequence in the adsorption reaction. However, a recent study using a synthetic decapeptide corresponding to the last 10 residue hydrophilic sequence at the C-terminus [21], demonstrated that the partial sequence in itself has only a very low adsorption affinity (of the same order of magnitude as those of single amino acids, such as aspartic acid and phosphoserine) and, as expected by its low adsorption affinity, shows a weak inhibitory activity on the growth of apatite crystals in dilute supersaturated solutions. Similar results were obtained using other amelogenin fragments [34]. Furthermore, previous studies [32] showed that all charge families of the 25 kd amelogenin adsorbed onto the crystals, indicating that the small difference in net electric charges among the 25 kd isomorphs do not significantly affect their adsorption affinity onto apatitic surfaces. Based on the foregoing results and consideration, our current conclusion is that no segmental sequences are enough to explain the properties of the intact amelogenin, and that the properties of the amelogenins (and their functional roles in enamel mineralization) may be determined by the whole molecular structure, including both the highly conserved segments at the N- and C-termini. It is also deemed likely that the molecular structure of the amelogenin in solution may change substantially by the cleavage of, first, the C-terminal hydrophilic segment and then the subsequent cleavage of the N-terminus; our studies on porcine amelogenins using NMR gave supportive evidence for this concept [31].

The overall results suggest that the nature and properties of the biominerals is primarily determined by the solution composition or, more accurately, the driving force (DS) in the media. It is also probable that the kinetics of mineralization (and thereby the nature and properties of formed biominerals) at early stages of amelogenesis is modulated substantially by the presence of physiological inhibitors, e.g., organic matrix. The secreted intact amelogenins have different properties from those of the degraded amelogenins. It is an intriguing hypothesis that the behavior of amelogenins in situ and their postulated functions are most likely related to their unique molecular structure. There is still need for further investigations to elucidate what is the molecular conformation of amelogenins in solution and in the adsorbed state, how amelogenetic mineralization is regulated through protein-mineral interaction in situ, and what is the nature and property of enamel minerals formed at various stages of amelogenesis.

Acknowledgements. The author is grateful to Drs. M. Fukae, T. Tanabe, M. Shimizu, and E.C. Moreno for their stimulating discussion and collaboration in conducting the present work. He is also grateful to Drs. S. Shimoda and Y. Miake for their technical help. This work was supported by grants DE07623 and DE08670 from the National Institute of Dental Research.

References

1. Suga S (1983) Comparative histology of progressive mineralization pattern of developing enamel. In: Suga S (ed) Mechanisms of tooth enamel formation. Quintessence, Tokyo, pp 167–203
2. Robinson C, Fuchs P, Deutsch D, Weatherell JA (1978) Four chemically distinct stages in developing enamel from bovine incisor teeth. Caries Res 12:1–11
3. Warshawsky H (1984) Ultrastructural studies on amelogenesis. In: Butler WT (ed) The chemistry and biology of mineralized tissues. Birmingham, Ala, pp 33–45
4. Takano Y, Hanawa M, Yamamoto T, Domon T, Fujinami H, Hanaizumi Y, Wakita M (1990) Time-related changes in the distribution of ^{45}Ca in the developing enamel of rat incisors as revealed by radioautography. J Biol Buccale 18:135–147
5. Bawden JW, Merriff DH, Deaton TG (1981) In vitro study of calcium-45 and phosphorus-32 uptake in developing rat molar enamel using quantitative methods. Arch Oral Biol 26:477–482
6. Termine JD, Belcourt AB, Christner PJ, Conn KM, Nylen MV (1980) Properties of dissociatively extracted fetal tooth matrix proteins. I. Principal molecular species in developing bovine enamel. J Biol Chem 255:9760–9768
7. Fincham AG, Belcourt AB, Lyaruu DM, Termine JD (1982) Comparative protein biochemistry of developing dental enamel matrix from five mammalian species. Calcif Tissue Int 34:182–189
8. Shimizu M and Fukae M (1983) Enamel Proteins. In: Suga S (ed) Mechanisms of tooth enamel formation. Quintessence, Tokyo, pp 125–141
9. Shimokawa H, Sobel ME, Sasaki S, Termine JD, Young MF (1987) Heterogeneity of amelogenin mRNA in the bovine tooth germ. J Biol Chem 262:4042–4047
10. Limeback H, Sakarya H, Chu W, Mackinnon M (1989) Serum albumin and its acid hydrolysis peptites dominate preparations mineral-bound enamel proteins. J Bone Mineral Res 4:235–241
11. Shimokawa H, Tamura M, Ibaraki K, Sasaki S (1989) Human amelogenin gene. In: Fearnhead RW (ed) Tooth enamel V. Florence, Yokohama, pp 301–305
12. Snead ML, Lau EC, Zeichner-David M, Fincham AG, Woo SLC, Slavkin HC (1985) DNA sequence for cloned cDNA murine amelogenin reveals the amino acid sequence for enamel-specific protein, Biochem Biophys Res Commun 129:812–818
13. Yamakoshi Y, Tanabe T, Fukae M, Shimizu M (1989) Amino acid sequence of porcine 25 kDa amelogenin. In: Fearnhead RW (ed) Tooth enamel V. Florence, Yokohama, pp 314–318
14. Weiner S (1986) Organization of extracellularly mineralized tissues: A comparative study of biological crystal growth. CRC Crit Rev Biochem 20:365–408
15. Linde A, Lussi A, Crenshaw MA (1989) Mineral induction by immobilized polyanionic proteins. Calcif Tissue Int 44:286–295
16. Herold R, Rosenbloom J, Granovsky M (1989) Phylogenetic distribution of enamel proteins: Immunohistochemical localization with monoclonal antibodies indicates the evolutionary appearance of enamelins prior to amelogenins. Calcif Tissue Int 45:88–94
17. Slavkin HC, Bessem C, Bringas P Jr, Zeichner-David M, Nanci A, Snead ML (1988) Sequential expression and differential function of multiple enamel proteins during fetal, neonatal, and early postnatal stages of mouse molar organogenesis. Differentiation 37:26–39
18. Aoba T, Tanabe T, Moreno EC (1987) Function of amelogenins in porcine enamel mineralization during the secretory stage of amelogenesis. Adv Dent Res 1:252–260

19. Aoba T, Moreno EC (1987) The enamel fluid in the early secretory stage of porcine amelogenesis: Chemical composition and saturation with respect to enamel mineral. Calcif Tissue Int 41:86–94
20. Aoba T, Fukae M, Tanabe T, Shimizu M, Moreno EC (1987) Selective adsorption of porcine amelogenins onto hydroxyapatite and their inhibitory activity on seeded crystal growth of hydroxyapatite. Calcif Tissue Int 41:281–289
21. Aoba T, Moreno EC, Kresak M, Tanabe T (1989) Possible roles of partial sequences at N- and C-termini of amelogenin in protein-enamel mineral interaction. J Dent Res 68:1331–1336
22. Aoba T, Moreno EC (1989) Mechanism of amelogenetic mineralization in minipig secretory enamel. In: Fearnhead RW (ed) Tooth enamel V. Florence, Yokohama, pp 163–167
23. Aoba T, Moreno EC (1992) Changes in the solubility of enamel mineral at various stages of porcine amelogenesis. Calcif Tissue Int 50:266–272
24. Moreno EC, Aoba T (1987) Calcium binding in enamel fluid and driving force for enamel mineralization in the secretory stage of amelogenesis. Adv Dent Res 1:245–251
25. Drinkard C, Gibson L, Crenshaw MA, Bawden JW (1981) Calcium binding by organic matrix of developing bovine enamel. Arch Oral Biol 26:483–485
26. Nelson DGA, Barry JC, Shields CP, Glena R, Featherstone JDB (1989) Crystal morphology, composition, and dissolution behavior of carbonated apatites prepared at controlled pH and temperature. J Colloid Interface Sci 130:467–479
27. Shimoda S, Aoba T, Moreno EC, Miake Y (1990) Effect of solution composition on morphological and structural features of carbonated calcium apatites. J Dent Res 69:1731–1740
28. Fukae M, Tanabe T (1987) Nonamelogenin components of porcine enamel in the protein fraction free from the enamel crystals. Calcif Tissue Int 40:286–293
29. Tanabe T, Aoba T, Moreno EC, Fukae M, Shimizu M (1990) Properties of phosphorylated 32 kD nonamelogenin proteins isolated from porcine secretory enamel. Calcif Tissue Int 46:205–215
30. Aoba T, Shimoda S, Akita H, Holmberg C, Taubman MA (1992) Anti-peptide antibodies reactive with epitopic domains of porcine amelogenins at the C-terminus. Arch Oral Biol 37:249–255
31. Aoba T, Kawano K, Moreno EC (1990) Molecular conformation of porcine amelogenins and its significance in protein-mineral interaction: 1H-NMR photo CIDNP study. J Biol Buccale 18:189–194
32. Tanabe T, Aoba T, Moreno EC (1988) Effect of fluoride in the apatitic lattice on adsorption of enamel proteins onto calcium apatites. J Dent Res 67:536–542
33. Aoba T, Collins J, Moreno EC (1989) Possible function of matrix proteins in fluoride incorporation into enamel mineral during porcine amelogenesis. J Dent Res 68:1162–1168
34. Aoba T, Moreno EC (1991) Structural relationship of amelogenin proteins to their regulatory function of enamel mineralization. In: Surface reactive peptides and polymers. American Chemical Society, Washington DC, pp 85–106

Recent Observations on the Formation of Carbonatoapatite In Vivo

TAKAAKI AOBA and EDGARD C. MORENO[1]

Summary. Biominerals in tooth and bone are most adequately described as carbonatoapatites. In order to elucidate the mechanism of carbonated apatite formation in vivo, our experimental strategies have been to (1) determine changes in the nature and properties of the biominerals with advancement of mineralization or tissue development, (2) assess the driving force for precipitation of the crystals in extracellular fluids, and (3) investigate temporal and spacial modifications of mineralization kinetics caused by various native regulators. Enamel mineralization taking place during amelogenesis is a unique and most useful model to investigate carbonatoapatite formation in vivo and the mineral changes at various stages of development. It was established that (1) substantial amounts of carbonate substitute both the PO_4^{3-} and OH^- ions in enamel crystals, (2) early enamel mineralization occurs in an environment isolated from circulating blood plasma, under a partial pressure of CO_2 of around 0.018 atm, and (3) the precipitation and growth of the crystals proceed steadily, without interference due to remodeling of the formed tissue, yielding the largest crystallites in the body. In the past few years, our effort was focused on the assessment of the labile and stable pools of the crystal constituents (i.e., carbonate, acid phosphate and fluoride) in developing enamel tissues, the development of a new solubility model taking into account most of the lattice constituents, and the determination of the solubility of enamel mineral under controlled partial pressures of CO_2. Although our knowledge is still limited and further refinement of the adopted stoichiometry and solubility models is required, the information obtained is valuable to predict the driving force for carbonatoapatite precipitation, as well as the stability of enamel mineral, in normal and pathologic environments.

Key words: Biomineralization — Amelogenesis — Carbonatoapatite — Stoichiometry — Solubility—Crystal growth

Introduction

Numerous reports in the literature have provided evidence that carbonation of calcium apatites is a common phenomenon in hard tissues. It is generally considered that most of the carbonate occupies sites in the crystalline lattice,

[1] Forsyth Dental Center, 140 Fenway, Boston, MA 02115, USA

although some of the ions may exist as CO_3^{2-} or HCO_3^- adsorbed on the crystal surfaces. The carbonate incorporation into the crystalline lattice occurs at 2 sites, originating 2 types of substitution [1–4]; in type B, the carbonate substitutes for phosphate, the planar CO_3^{2-} molecule occupying one of the planes in the PO_4^{3-} tetrahedron; in type A, CO_3^{2-} replaces the hydroxyl ion. Previous studies using polarized infrared (IR) spectroscopy [2] demonstrated that the type B substitution gives rise to distinct absorption bands at 1465, 1412, and 873 cm^{-1}, and bands at 1545, 1450, and 880 cm^{-1} correspond to the type A substitution. Recent studies by Fourier transform infrared spectroscopy (FTIR), in conjunction with resolution-enhancing techniques [5, 6], distinguished an additional CO_3^{2-} band around 865 cm^{-1} in the spectra of bone and enamel minerals, which was ascribed to the labile carbonate on the crystal surface.

Concerning the effect of carbonate ions on precipitation of apatite crystals, it was shown that carbonate ions in highly supersaturated solutions inhibit apatite formation from precursors [7, 8] and also inhibit the growth of apatite crystals along the *c*-axis direction [9], thus affecting the morphology of the grown crystals [10]. It was also reported that the carbonate substitution in apatite crystals causes an increase in the number of lattice defects [11, 12], increasing the solubility of the crystals [13–15]. However, little is known about the thermodynamic properties of carbonated biominerals and the mechanism of their precipitation and growth under normal and pathological conditions. Thus, our interest in the past few years has been to investigate the nature and properties of carbonated bioapatites, in particular enamel mineral formed at various stages of amelogenesis.

Enamel Mineralization as a Unique Model for Biological Carbonatoapatite Formation

Carbonate is one of the prominent constituents (3 wt% or more) of enamel mineral [16, 17]. Recent studies by Elliott et al. [18] demonstrated that approximately 10% to 15% of the carbonate ions in mature human enamel is in the OH^- position, while the rest substitute for PO_4^{3-} ions. Most likely, these carbonate substitutions in enamel crystals take place when the mineral precipitates during amelogenesis. In fact, the relative enrichment of carbonate in early developing enamel tissue, as well as the association of carbonate with the forming enamel crystals at subsequent stages, has been reported for various mammalian species [19, 20]. There was a paucity of information, however, about the microenvironment surrounding the precipitating biominerals. In Table 1 are shown available data concerning electrolyte concentrations in various extracellular fluids relevant to mineralization in mammalian species, as well as the corresponding values for the blood plasma. Using a micropuncture procedure, Howell et al. [21] aspirated fluid samples from rat hypertrophic (normal and rachitic) cartilage and analytically determined the total concentrations of major electrolytes. Wuthier [22] also determined the electrolyte

Table 1. Electrolyte concentrations of extracellular fluids

Sample	*[Ca]* *mM*	*[total P]* *mM*	*[Mg]* *mM*	*[Na]* *mM*	*[K]* *mM*	*[Cl]* *mM*	*[HCO_3]* *mM*	*pH*
Hypertrophic cartilage fluid								
Rat normal[a]	1.5	2.6					36	7.58
Rat rachitic[a]	1.8	1.6					31	7.52
Chicken normal[b]	3.1	1.5	0.91	115	15	111		7.49
Bone (bovine) fluid[c]	0.48	1.8	0.4	125	25	130		
Enamel (pig) fluid[d]	0.50	3.9	0.83	139	21	146	10	7.26
Blood plasma								
Rat[a]	2.3	2.8					22	7.39
Chicken[b]	3.2	1.8	1.1	156	11	118		7.35
Bovine[c]	1.5	1.8	0.7	140	4	100		
Pig[d]	2.9	2.4	1.1	156	12	168		7.26

[a] Howell et al. [21]; [b] Wuthier [22]; [c] Neuman [23]; [d] Aoba and Moreno [24]

concentrations in chick cartilage fluid; his data were corrected by taking into account possible leakages of electrolytes from the surrounding cells. Neuman [23] assessed concentrations of electrolyte constituents in bovine "bone fluid" and compared them with the corresponding values in blood plasma. Recently, we succeeded in separating the liquid phase from the secretory enamel of porcine teeth by ultracentrifugation and determined the concentrations (and activities) of most of the lattice ionic constituents [24]. Examination of the data shown in Table 1 leads to the conclusion that (1) mineralization in cartilage, bone, and enamel occurs in fluid environments having a composition different from that of circulating blood, (2) all mineralizing systems contain sufficient carbonate ions to be incorporated into the crystal lattice, and (3) the CO_3 concentrations in these fluids are always higher than the concentrations of coexisting Ca and phosphate ions. More important, the composition of the enamel fluid is significantly different from those of the cartilage and bone fluids, specifically with respect to concentrations of calcium, phosphate and carbonate, and pH value. This difference indicates that enamel mineralization in the early developmental stages occurs in an isolated fluid environment.

In addition to the uniqueness of enamel fluid composition, mammalian amelogenesis provides an excellent model for the investigation of the carbonated hydroxyapatite formation, since the mineralization process is not intervened by any remodeling of the formed tissue, as in the case of bone (and cartilage) mineralization. Furthermore, differing from the rapid deposition of dentin crystals at the mineralization front, the growth of enamel crystals proceeds gradually, but to a significant extent, through all the developmental stages, resulting in the largest carbonatoapatite crystals in the body. An important conclusion drawn from ultrastructural studies [25–27] is that the advancement of enamel mineralization is not due to an increase in the number but in the size of the crystallites (i.e., crystal growth). Our recent studies using high resolution electron microscopy [28] showed that the enamel mineralization process is characterized by the initial precipitation of thin ribbons, the non-uniform accretion of apatite unit cells on prism planes, and the formation of flattened hexagonal crystals as revealed by cross-sectional images. Although not yet defined completely, it appears that the obtained morphology of enamel crystals and their growth kinetics at various developmental stages are controlled temporally by the modulation of cellular (ameloblast and surrounding enamel organ cells) activities and the presence of various regulators of apatite crystal growth (e.g., matrix proteins, F^-, and Mg^{2+}).

In general, the forming enamel during amelogenesis is divided into secretory, early maturing (including transitional), and late maturing (hard mature) stages on the basis of mineralization progress and associated alterations in proteinaceous matrix and tissue structures [29, 30]. Rodent incisors, growing and erupting continuously, have been widely used by many investigators, since a full spectrum of amelogenesis can be observed from the apical through the cuspal ends in a single tooth. However, it is of importance to realize that enamel mineralization during amelogenesis advances in 2 directions: from the outer to the inner zones as well as from the apical to the cuspal ends [31]. In this respect, porcine enamel is a suitable model for investigation of carbonatoapatite formation during amelogenesis, since the large tooth size facilitates dissection and recovery of sufficient amounts of enamel samples at various developmental stages, i.e., the outer (younger) secretory, inner (older) secretory, early maturing, and mature tissues. Indeed, substantial changes in the crystal morphology and protein constituents were observed between the outer (close to the ameloblasts) and inner zones of secretory porcine enamel [32]. It is also known that the chronology of enamel mineralization (and the tooth shape) of this animal is similar to that of human enamel [33]. Thus, most of our work cited in this report, i.e., changes in chemical composition [20, 32, 34, 35], sites of carbonate substitution [20, 28], morphology of enamel crystals [28, 32] and their solubility [36], was conducted with enamel samples dissected from unerupted permanent incisors of piglets.

Changes in Chemical Composition of Enamel Mineral During Amelogenesis

As mentioned above, much work has been conducted on the composition of the inorganic constituents of enamel over the last three decades. However, there remains a paucity of information about the state or environment of ionic constituents in enamel tissues at various developmental stages. Specific questions to be addressed are: (1) whether part of the ionic constituents of enamel tissues may be in labile forms (i.e., organically bound or adsorbed on crystal surfaces) rather than in a stable form (i.e., incorporated into the crystal lattice), (2) whether protonated ionic species, e.g., HCO_3^- and HPO_4^{2-}, exist in significant quantities in enamel crystals, and, as a consequence, (3) what are the true lattice stoichiometries of enamel crystals formed at various stages of enamel mineralization. The fluoride uptake into developing enamel has been studied by many investigators [37, 38] since it was disclosed that the incorporated fluoride ions are effective stabilizers of enamel mineral, whereas, at high concentrations, fluoride induces defective enamel structures. However, there still remain controversies about the state of the fluoride in developing enamel tissues or enamel proteins-fluoride binding [39, 40]. Thus, the contents of calcium, total phosphate, carbonate, acid phosphate, and fluoride were determined in the enamel samples dissected from regions of porcine incisors corresponding to the various developmental stages. Significant attention was given to assessment of the labile pools of carbonate (removal of surface CO_3 species by plasma ashing in an activated oxygen atmosphere at about 60°C [20]), acid phosphate (successive extraction with 3 mM phosphate solutions at pH 11 [35]), and fluoride (successive extractions with neutral NaCl solutions and carbonate buffers [34]).

Table 2 shows the results of chemical analyses of the total and the labile pools of carbonate, acid phosphate, and fluoride. Both the calcium and phosphorus contents (% dry weight) in porcine enamel tissue increased consistently with tissue development, from the outer secretory (Ca, 15%; P, 8%) to the

Table 2. Chemical composition of porcine enamel samples obtained at various developmental stages

Samples	*Ca* *wt%*	*P* *wt%*	*CO_3* *wt%*	*% P* *as HPO_4^{2-}*	*F* *wt% $\times 10^2$*
Outer secretory enamel	22.8	11.6	3.8 (18)	15.6 (60)	1.0 (30)
Inner secretory enamel	25.5	12.8	4.0 (12)	10.8 (45)	0.5 (25)
Maturing enamel	30.8	15.2	3.8 (7)	8.3 (35)	1.5 (**)
Mature hard enamel	35.9	17.5	2.8 (**)	5.3 (5)	

** trace (<5%)
Figures in parenthesis represent the labile pool as percentages of the total CO_3^{2-}, HPO_4^{3-}, and F^- in each enamel tissue

mature enamel (Ca, 36%; P, 17.5%). In contrast, the total carbonate content decreased only modestly from the maximum content of 4.0 wt% in the inner secretory enamel to 2.8 wt% in the mature enamel. Importantly, significant portions (about 18% and 12%) of the carbonate in the outer and inner secretory enamel, respectively, were lost by low temperature ashing, whereas this lability of carbonate disappeared as the enamel tissue developed. The latter reduction in carbonate suggests that part of the carbonate is adsorbed on the crystal surfaces in the early stages of mineralization, while, with the advancement of mineralization, the carbonate, almost in its totality, becomes a lattice constituent in the enamel crystals. Moreover, if carbonation of enamel crystals is expressed by the stable pool of CO_3^{2-} per mol of Ca (or P), the carbonate incorporation in enamel mineral was revealed to be fairly constant (stable $CO_3/Ca = 0.09$) in the secretory enamel while it decreased from 0.07 in the maturing to 0.05 in the mature enamel. Most likely, the foregoing changes in carbonate substitution reflect compositional changes in the enamel fluid between the secretory and the further advanced stages (presumably related to the activity of the ameloblasts).

Concerning the acid phosphate content of enamel samples, it was found that (1) the outer secretory enamel contained the highest HPO_4^{2-}, corresponding to 15% or more of the total phosphate, (2) the HPO_4^{2-} content decreased with tissue development, and (3) the HPO_4^{2-} content in the mature enamel was about 5.3% of the total phosphate. The maximum value of the HPO_4^{2-} obtained thus far from the dissected porcine enamel corresponds to one-half of the content anticipated from the stoichiometry of octacalcium phosphate, $Ca_4H(PO_4)_3 \cdot 5H_2O$. Interestingly, a substantial part (60%) of the total acid phosphate in the outer (young) secretory enamel was in labile forms (i.e., lost after the extraction with the alkaline solution, due to deprotonation or exchange with PO_4^{3-} in solution [35]), whereas the corresponding pools decreased substantially to 45% in the inner (older) secretory enamel, to 35% in the maturing enamel, and to only a marginal fraction (<5%) in the mature enamel. On the basis of electron microscopic observations of enamel crystal growth [28, 32], it seems reasonable to think that the decrease of the total and the labile HPO_4^{2-} with the advancement of mineralization may result from the

Table 3. Chemical composition of rat secretory enamel from animals ingesting fluoride

Animal group	*Ca* *wt%*	*P* *wt%*	*F* *wt%* $\times 10^2$
Control	17.1	8.8	1.2 (23)
25 ppm	17.8	9.0	17.5 (**)
50 ppm	17.5	9.0	31.1 (**)
75 ppm	18.0	9.1	40.9 (**)
100 ppm	17.7	9.1	54.5 (**)

** trace (<5%)
Figures in parenthesis represent the labile pool as percentages of the total F^- in each enamel tissue

burying of an initially precipitated domain rich in HPO_4^{2-} with a subsequent deposition of an apatitic phase having different stoichiometries from that of the initially precipitated thin ribbons.

We also determined the fluoride content in porcine secretory and mature enamel samples with special attention to possible association of fluoride with the matrix proteins. Most of the amelogenins in the enamel tissue were extracted with a carbonate buffer at pH 10.8. The fluoride in the original and treated enamel samples was measured by the microdiffusion procedure, in conjunction with an ion-selective electrode [34]. The results (Table 2) indicate that (1) the total F per unit mass of Ca (or P) in the tissue was maximal at the beginning of the secretory stage and decreased (and leveled off) with the advancement of mineralization, and (2) interestingly enough, part of the fluoride in the secretory enamel, about 25%–30% of the total, was removed with the extraction of organic matter, mostly amelogenins. These results suggest that a fraction of the fluoride in the secretory enamel may be associated with the organic matter. However, the nature of that association remains obscure.

A question of interest in understanding the role of fluoride in enamel mineralization is whether the labile fluoride is simply a pool available for incorporation into enamel crystals, or if it may have other specific functions, e.g., modification of protein properties or their degradation processes in the process of enamel mineralization. A partial answer to this question was obtained in studies on the uptake of systemic fluoride by rat enamel [43]. The animals were given either deionized water or water containing 25, 50, 75, and 100 ppm fluoride as NaF. Table 3 shows the Ca, total P, and fluoride contents of the secretory enamel obtained for each of the animal groups. The results clearly showed that, whereas there were no significant differences in the Ca and P contents among the animal groups, the total fluoride content of the enamel tissue increased with increasing the concentration of fluoride in the drinking water. A conspicuous finding was that no appreciable increase in the labile fluoride was obtained in any of the animal groups. In parallel studies, it was verified that the systemic fluoride ingestion at the doses tested did not cause discernible changes in the properties of amelogenins and their degradation cascade during the secretory stage. Furthermore, in vitro crystal growth studies yielded evidence that coating hydroxyapatite seeds with enamel matrix proteins resulted in a retardation of fluoride incorporation, as a result of an inhibition of apatite precipitation. Based on the foregoing experimental data, it seems that the fluoride incorporation into enamel crystals is intimately regulated by the kinetics of mineralization but, once an excess of fluoride ions beyond the magnitude of the labile pool (depending on the amounts and properties of matrix proteins) becomes available under relatively high fluoride regimes, the excess F^- promotes the precipitation reaction, as demonstrated in vitro [44], leading to the accretion of fluoride in the forming crystals.

Structural Changes of Enamel Mineral at Early Stages of Amelogenesis

The same batches of porcine enamel samples used for the chemical analysis were examined by X-ray diffraction and FTIR spectroscopy. Figure 1 shows typical X-ray diffraction patterns recorded from the porcine enamel samples. The observed reflections, although most of them in the outer secretory enamel samples were broadened (reflecting a low degree of crystallinity of the initially precipitated mineral), match those for the apatite lattice. Table 4 shows the lattice parameters and relative crystallinity of enamel mineral. Developmental advancement from the outer secretory to the inner secretory as well as early maturation stages brought about significant changes in the unit cell parameters, namely, shrinkage and expansion of the *c* and *a* unit cell dimensions, respectively, and an improvement of the crystallinity, particularly in the *a*-axis direction. These changes in the unit cell parameters are consistent with the relative decrease of the carbonate and acid phosphate contents and the changes in CO_3^{2-} substitution sites mentioned above. Also, the increasing crystallinity is consistent with the observation that the advancement of enamel mineralization is characterized by an increase in the size of the crystallites.

For comparison, in Fig. 2 are shown typical X-ray diffraction patterns of synthetic carbonatoapatites obtained at 37°, 60°, and 80°C. It is pertinent to

Table 4. Lattice parameters and line broadening of porcine enamel apatites obtained at various developmental stages

Sample	*Lattice parameter (Å)*		*Line broadening (°(2θ))*	
	c	*a*	*(002)*	*(310)*
Outer secretory enamel	6.890	9.433	0.38	1.14
Inner secretory enamel	6.886	9.437	0.30	0.90
Maturing chalk-like enamel	6.884	9.440	0.28	0.64
Mature hard enamel	6.882	9.443	0.24	0.39

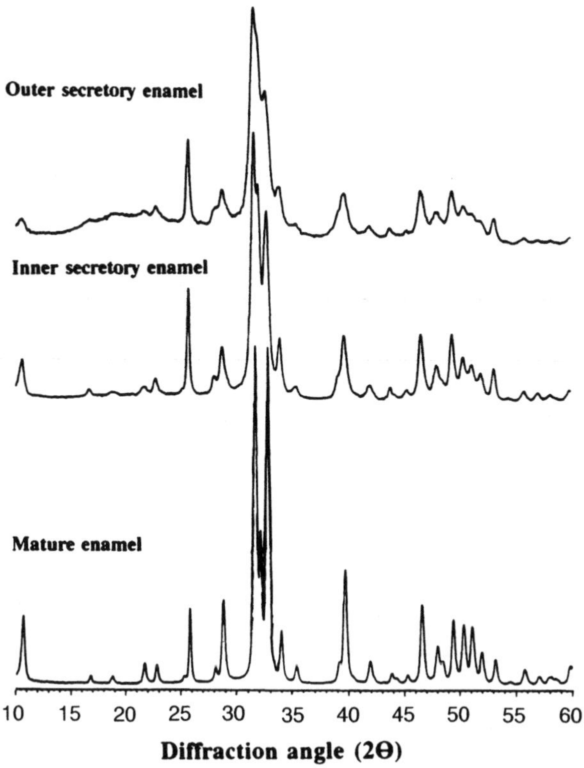

Fig. 1. X-ray diffraction patterns of porcine enamel at various developmental stages

note that the broadened pattern characteristic of the initial enamel mineral was similar to those for bone and dentin minerals and to the pattern of carbonatoapatite synthesized in aqueous solutions at 37°C [11]. In contrast, the pattern obtained for mature enamel was well resolved and comparable to those of carbonatoapatites prepared at higher temperatures [9].

Large fluoridated apatite crystals can be precipitated at 37°C in the presence of high concentrations of fluoride (1 to 10 ppm) [45]. However, it remains an enigma how mammalian amelogenesis yields large enamel crystals at the physiological temperature, with a relatively low concentration of fluoride in the enamel fluid. Presumably, one key mechanism in enamel mineralization might be the prevention of proliferation of nuclei, thus facilitating the growth of a limited number of the crystals over prolonged periods (several weeks to months). Putative regulators that may play this inhibitory role are the secreted matrix proteins. At present, it remains an intriguing and important challenge to synthesize carbonated hydroxyapatites having crystal sizes comparable to that of enamel crystals, using well-defined systems at the physiological temperature.

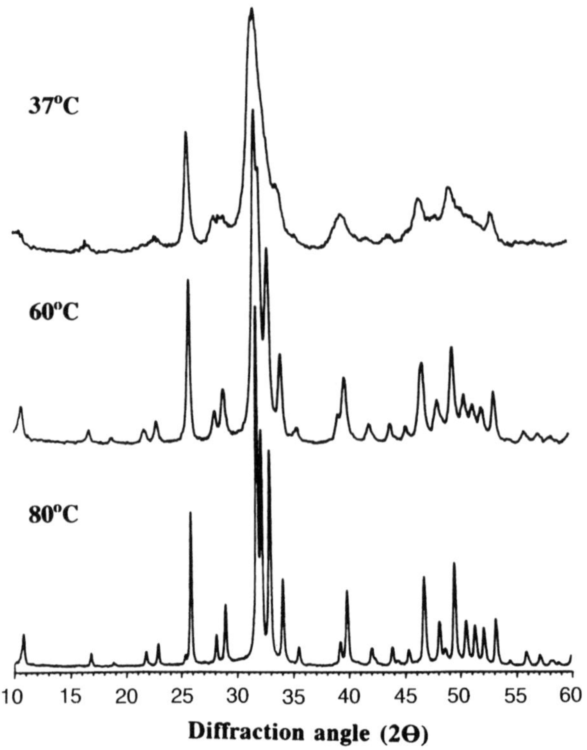

Fig. 2. X-ray diffraction patterns of carbonatoapatites precipitated at various temperatures

Figure 3 shows typical FTIR spectra obtained for the secretory and mature porcine enamel. Broad peaks at 1660, 1530, and 1450 cm^{-1} observed for the secretory enamel sample originate from the amide groups of the organic matter. These bands became obscure in the spectra obtained for the mature enamel, since little organic matter remained in the tissue. In both enamel samples, CO_3^{2-} bands were detected at 1417 and 1467 cm^{-1} corresponding to the B-type substitution. As reported previously [20], analysis of the mu$_2$ CO_3^{2-} band at 880–870 cm^{-1} also confirmed that the carbonate in enamel apatites formed in the early stages of porcine amelogenesis substitutes mainly for phosphate ions, while, with developmental advancement, there is an increase in the amount of CO_3^{2-} occupying OH^- sites in the crystalline lattice. Importantly, a comparison of both spectra in Fig. 3 revealed that the peak-maxima of mu$_3$ PO_4^{3-} bands shifted from the 1031 cm^{-1} region in the secretory enamel to the 1040 cm^{-1} in the maturing sample. This observation might be consistent with an OCP-like mineral formation at the onset of mineralization because the corresponding bands in synthetic OCP appear at 1025 and 1035 cm^{-1} [46]. In this respect, it is interesting that similar shifts in the mu$_3$

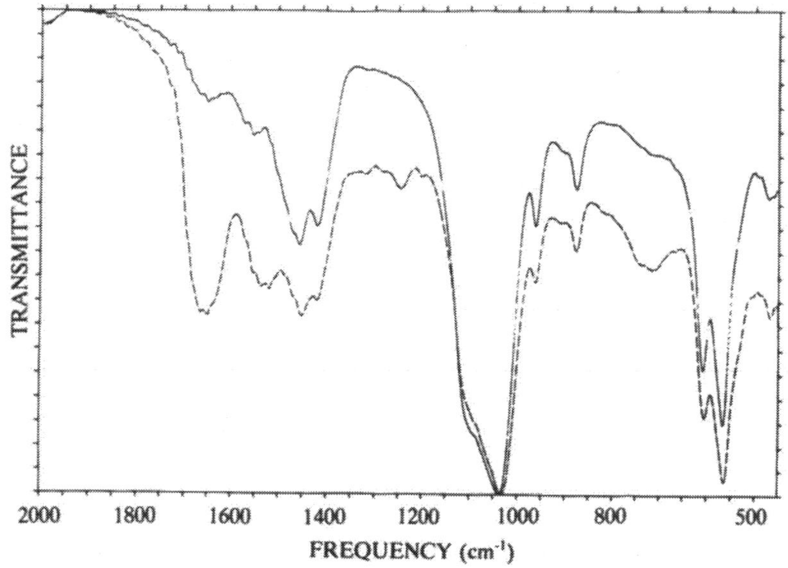

Fig. 3. FTIR spectra of secretory and mature porcine enamel. Note the shift of mu_3 PO_4^{3-} bands from 1031 to 1040 cm^{-1} with tissue development

PO_4^{3-} bands were observed with ageing of the mineral deposited in matrix vesicles [47] and deposits of calcifying cartilage [48].

Driving Force for Precipitation and Growth of Apatite Crystals

Previous work about crystal growth of hydroxyapatite [44, 49] demonstrated that the nature of the precipitating phase is largely dependent on the experimental conditions (e.g., degree of solution supersaturation, concentration of crystalline seeds, temperature, existing regulators, etc.). In in vivo mineralization, the degree of supersaturation (DS) is of particular importance. In the case of calcium phosphates, the DS can be expressed as the ratio of the ionic activity product (IP) of the lattice constituents in solution to the solubility product (Ksp) of the solid concerned, DS = IP/Ksp or in a form of mean activities $DS = (IP/Ksp)^{1/t}$, where t is the total sum of ions that the compound generates upon solution. In testing the dependence of the growth rate of hydroxyapatite on the DS of experimental solutions, a series of kinetic studies were conducted at 37°C with a batch of synthetic hydroxyapatite crystals (6.3 m^2/g) as seeds [50]. In those studies, the initial composition of the solution used was 1.06 mM Ca, 0.63 mM P, 50 mM NaCl with a pH value of 7.4. The DS value with respect to hydroxyapatite is 16.1. The composition ([Ca], [P] and pH) of the solution started changing immediately upon addition of the

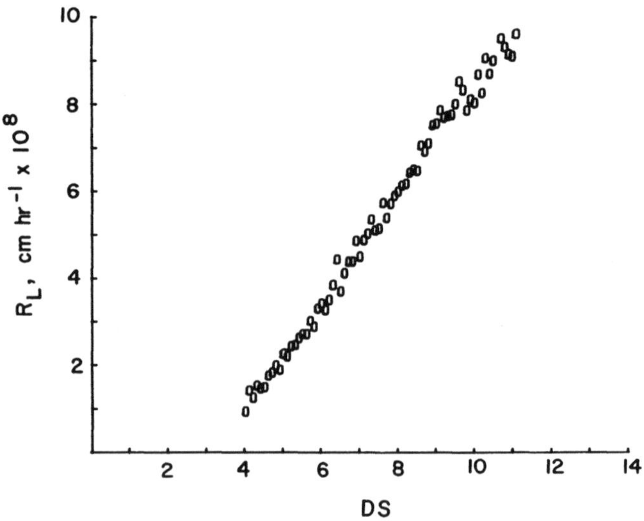

Fig. 4. Linear rate (R_L) of crystal growth of hydroxyapatite at 37°C as a function of the degree of supersaturation (*DS*) in solution with respect to hydroxyapatite

seeds (0.1 to $0.3\,m^2/100\,mL$) at time zero and the magnitude of these changes were related to the degree of supersaturation DS. The linear growth rate, R_L, was determined from the volume of HA precipitated at any given time calculated on the basis of the phosphorus (or Ca) precipitated (using a density of $3.1\,g/cm^3$ for the precipitated HA); the mean linear growth up to that time is V/S where S is the surface area of the added seeds. Values of V/S were obtained for each of the experimental times. Then, a polynomial of the third degree in time was fitted to the values V/S. The linear growth rate for each experimental time was then obtained by differentiating the polynomial and using the numerical value of time in the resulting expression. In Fig. 4 are plotted the values of R_L versus their corresponding values of DS for a number of experimental systems. In this plot, the DS value was expressed by the mean activities $DS = (IP/Ksp)^{1/9}$. The results clearly showed that the growth rate of hydroxyapatite increases with increasing of the DS. The obtained relationship between the growth rates and the driving force implies a plausible mechanism for the crystal growth process [44], although further discussion in this respect is beyond the scope of this report.

In independent experiments, we also investigated the dependence of the precipitation kinetics on surface areas of the apatite seed crystals [50] and the effects of various native regulators on the growth kinetics and the nature of the precipitating crystals [51]. In Fig. 5 are shown typical results on the effects that fluoride and magnesium ions have had on crystal growth in dilute supersaturated solutions having compositions similar to that of the enamel fluid. The initial pH of the supersaturated solution was 7.3 and the initial concentrations of Ca and total P were 0.5 mM and 3.0 mM, respectively. The ionic strength

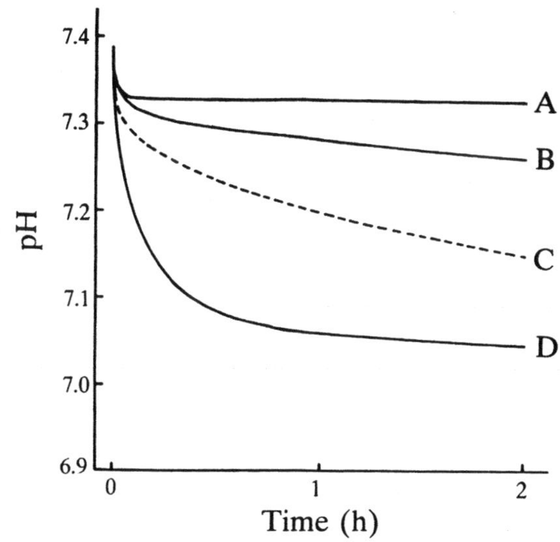

Fig. 5. Effects of fluoride and magnesium ions on the precipitation kinetics of hydroxyapatite at 37°C. Concentrations of the additives in the initial experimental solution were: *A*, [Mg] = 1 mM; *B*, [Mg] = 1 mM and [F] = 1 ppm; *C*, control ([Mg] = 0 mM and [F] = 0 ppm); *D*, [F] = 1 ppm

Table 5. Inhibition of hydroxyapatite crystal growth due to porcine amelogenins

Amelogenin MW	*Total amounts of proteins used mg*	*Amounts adsorbed onto crystals mg (mg/m²)*	*% Inhibition of initial growth rate*
25 kd	0.33	0.26 (0.82)	22
	0.66	0.47 (1.57)	33
20 kd	0.42	0.03 (0.09)	<5
	0.64	0.04 (0.13)	5

of the solution was maintained by using 160 mM NaCl as a background electrolyte. The experimental solution was stable (no spontaneous precipitation) at 37°C over 5 h. Comparison of the plot of solution pH versus experimental time for the control run (seeding in the absence of additives) with the other curves, shows that the addition of either fluoride (1 ppm) or magnesium (1 mM) or both to the experimental solution significantly affects (acceleration by F^- and partial inhibition by Mg^{2+}) the kinetics of precipitation onto the seed crystals.

Inhibitory effects on the apatitic precipitation were also observed when the seed crystals were pretreated with the amelogenin proteins. The proteins tested were porcine 25 kd amelogenin (intact secreted protein) and its degraded product (20 kd amelogenin). The seed crystals were pretreated with specified amounts of individual proteins, as reported previously in detail [52]. Table 5 shows the effect that pretreatment of seed crystals has on the precipitation kinetics of apatite. The inhibition on the crystal growth of hydroxyapatite due to the 25 kd and 20 kd amelogenins was quantified in terms of the resulting reduction in initial precipitation rate, (Rc − Ri)/Rc × 100, where Rc and Ri were the initial precipitation rates determined using untreated and pretreated seeds with these proteins, respectively. It is apparent that (1) the 25 kd amelogenin yielded a higher inhibitory activity than the 20 kd protein, and (2) the observed inhibitory activity is related to the adsorption of the moieties onto the seed crystals. In relation to the possible inhibitory function of the intact amelogenin, it is of interest to note that the secreted 25 kd amelogenin was concentrated in the outermost layer of secretory enamel, but its amount decreased substantially in the second and underlying inner layers [32], i.e., in the older tissue. This localization of the intact protein in the forming enamel, as well as its inhibitory activity obtained in in vitro systems, may be related to the observation [32] that the crystallites existing in the outermost layer were thin plate-like in appearance but the width or thickness of the crystallites was already increased in the second layer. The foregoing results lead to the concept that enamel mineralization, in the early stages of amelogenesis, is intimately related to the secretion and degradation, i.e., enzymatic cleavages of specific segments, of amelogenins [53, 54].

Solubility of Enamel Mineral Under Controlled Pco_2

As explained in the preceding paragraphs, knowledge about the solubility product (Ksp) of the forming biomineral is essential, not only for defining the thermodynamic stability of the enamel mineral at each stage of the developmental processes, but also for assessing the driving forces involved in the precipitation of the apatite crystals. In the literature, thermodynamic solubilities of hydroxyapatite have received extensive attention. However, no definite thermodynamic approach has been applied to carbonatoapatites (better prototypes for biominerals than hydroxyapatite) or carbonated biominerals. One of the reasons for the paucity of information in the study of the thermodynamic properties of carbonated biominerals is that experimentation and interpretation of the analytical results are more complicated in the $Ca(OH)_2$-CO_2-H_3PO_4-H_2O system than those in the ternary system $Ca(OH)_2$-H_3PO_4-H_2O, or in other four-component systems without a gaseous component. In addition, the stoichiometry of the solid concerned is complicated and not well established. In some studies [55, 56], the stoichiometry of hydroxyapatite was assigned to dental enamel, and Ksp values were calculated on such a basis. However, this approach (neglecting an important constituent, carbonate) is a gross approximation of questionable validity; the obtained values of the solubility product constants are only of limited value for predicting the stability of the enamel mineral or the driving force for mineralization in physiological media.

For the last few years, our interest was focused on assessment of the solubility of porcine enamel at various developmental stages under controlled partial pressures of CO_2 (Pco_2). Powdered enamel samples (100 mg/100 mL) were equilibrated in dilute H_3PO_4 solutions (0.01 to 1.1 mM) in atmospheres of N_2-CO_2 mixtures (1.86 and 1.75% in CO_2), which bracketed the Pco_2 value (1.8%) determined in the enamel fluid [53]. Equilibrium was assumed when the solution composition (pH, concentrations of Ca, P, Mg, Na, and K) did not change for a period of 3–4 days, a procedure which took from 15 to 25 days. Possible effects of any organic components remaining in the samples particularly in the secretory enamel, on the solution composition, e.g., due to formation of metal-complexes, were ruled out because, in most cases, the value of the calculated Ca^{2+} activity was in good agreement (less than 5% discrepancy) with the value determined using specific calcium ion and reference electrodes. Since analytical work on the enamel sample disclosed that acid phosphate was also an important constituent of the crystal lattice [35], the solubility data was interpreted on the basis of the given stoichiometry $Ca_{5-x}(HPO_4)_v(CO_3)_w(PO_4)_{3-x}(OH)_{1-x}$. The stoichiometric coefficients, x, v, and w, were calculated from the analytical values of Ca/P, HPO_4/P_{total} and CO_3/Ca molar ratios [36]. On the basis of the adopted stoichiometry, the equilibrium in solutions saturated with enamel mineral can be expressed by the following equation $(Ca^{2+})^{5-x}(HPO_4^{2-})^v(CO_3^{2-})^w(PO_4^{3-})^{3-x}(OH^-)^{1-x} = K_{EN}$, in which the parentheses indicate activities of the species enclosed and K_{EN} is the solubility product constant for enamel mineral. Table 6 shows the stoichiometries of enamel samples obtained at various developmental stages and their K_{EN} values, which were calculated from the individual solution composition at equilibrium. The results indicate that (1) the outer (younger) secretory mineral was the most soluble, and (2) the solubility of enamel mineral decreased with advancing developmental stages. As reported previously [24], it is reasonable to assume that the enamel fluid is in equilibrium with the enamel mineral formed in the secretory tissue. Using the values for the stoichiometric coefficients shown in Table 1, the values of IP in the secretory enamel fluid were calculated from its composition; they are shown in the last column of Table 6. Interestingly enough, the K_{EN} and IP values exhibit an excellent agreement for the most soluble mineral (outer secretory enamel), whereas the discrepancy between the values of K_{EN} and the IP became appreciable for the samples separated at the advanced developmental stages. This fact is indirect support of the concept that the ionic composition in the enamel fluid changes at various stages of amelogenesis, probably due to changes of the cellular metabolic activities.

The Mechanism of Carbonatoapatite Formation During Amelogenesis

All the experimental data and considerations mentioned above strongly suggest that carbonate incorporation in bioapatites is directly related to the thermodynamic and kinetic aspects of mineralization processes, which, in turn, are primarily determined by the DS of the fluid environment. An important observation, made in previous solubility studies with hydroxyapatite (and enamel mineral) under controlled Pco_2 [57], was that precipitation of a car-

Table 6. Stoichiometry and solubility of enamel mineral formed at various developmental stages

Sample	*Stoichiometry*	K_{EN}[a]	*IP in fluid*[b]
Outer secretory enamel	$[Ca]_{4.16}[HPO_4]_{0.41}[CO_3]_{0.42}[PO_4]_{2.16}[OH]_{0.18}$	6.33×10^{-41}	6.47×10^{-41}
Inner secretory enamel	$[Ca]_{4.29}[HPO_4]_{0.31}[CO_3]_{0.40}[PO_4]_{2.29}[OH]_{0.29}$	1.30×10^{-43}	4.63×10^{-43}
Maturing chalk-like enamel	$[Ca]_{4.45}[HPO_4]_{0.23}[CO_3]_{0.32}[PO_4]_{2.45}[OH]_{0.45}$	1.34×10^{-46}	3.31×10^{-46}
Mature hard enamel	$[Ca]_{4.63}[HPO_4]_{0.11}[CO_3]_{0.25}[PO_4]_{2.63}[OH]_{0.65}$	1.33×10^{-49}	8.08×10^{-49}

[a] K_{EN}, Solubility product constant determined at 25°C, based on the solution composition at equilibrium
[b] *IP in fluid*, Ionic activity product calculated with the analytical values of lattice constituents in the enamel fluid shown in Table 1

bonatoapatite occurred concomitantly with the dissolution of the original solid in acidic media, meaning that carbonatoapatite is the stable mineral phase in media having certain Pco_2. This observation, as well as the ubiquitous presence of carbonate ions in extracellular fluids shown in Table 1, provides an explanation for the fact that carbonation of apatite crystals is a common, not an exceptional, phenomenon in hard tissues. However, it is not yet clear how the magnitude and location of carbonate substitution in the crystal lattice are determined in the process of biomineralization.

In order to gain an insight into the mechanism determining carbonation of biominerals, we conducted a series of carbonatoapatite preparations at various temperatures and with solutions having wide ranges of composition [9]. The characterization of the products by FTIR, X-ray diffraction, and electron microscopy, lead to the conclusion that the magnitude and sites of carbonate substitution and the growth and morphology of the formed crystals are highly dependent on both the degree of supersaturation (with respect to calcium phosphates) and the presence of "regulators" (i.e., CO_3^{2-} and F^-) in solution. Specifically, low concentrations of carbonate in solution favored the A-type substitution in the precipitated crystals [9, 58]. With increasing carbonate content, the fraction of carbonate in the OH^- site decreased substantially, whereas the carbonate in the PO_4^{3-} sites became prominent. In the presence of high concentrations of fluoride, acicular apatite crystals were precipitated initially, growing into rod-like crystals having symmetric-hexagonal cross sections and the carbonate substitution was almost exclusively of the B-type. Based on the foregoing results, we hypothesized that the selective occupation of hydroxyl sites by carbonate at the late stage of enamel formation is due to a competition for two sites, one of which (A-type) is favored energetically but is present in a limited number in the system, whereas the second site (B-type) is present in a larger number. With lower availability of carbonate, the energetic considerations prevail but, at high concentrations, the probability of occupancy in each substitution site is essentially determined by their respective number of sites. The above consideration supports the connotation that the preferential B-type substitution found in enamel crystals at early developmental stages could indicate the formation of initial biomineral in the presence of relatively high carbonate and that the subsequent apatitic growth, incorporating more carbonate in the OH^- sites, occurs in media having lower Pco_2. Similar energetic considerations may apply to the occupancy of the OH^- sites by carbonate, hydroxyl, and fluoride ions. If a true competition exists among the ions, the fluoride that introduces the least lattice disturbance [59] should be most favored. Thus, it is not surprising that, even when the carbonate concentration in solution is several times higher than the fluoride concentration, the precipitated carbonatoapatite would have all its A-sites occupied by fluoride. This is most likely the case seen for the francollite formation in some fish enameloids [28, 60].

An alternative mechanism for the incorporation of carbonate in enamel crystals, involving the initial precipitation of OCP and its hydrolysis, has also been proposed [61, 62]. Chickerur et al. [63] demonstrated that the B-type substitution of carbonate, which is the only observable substitution in the youngest (outer secretory) enamel [20], takes place in the apatite crystal lattice when OCP crystals are hydrolyzed in the presence of Na^+ and CO_3^{2-} at 37°C. As reported previously [64], our computer simulation on the basis of enamel fluid composition suggested that the fluid may become supersaturated with respect to OCP with an increase in the concentration of total Ca of more than 1 mM. This situation is likely to occur in the vicinity of the ameloblasts, when the cells transport Ca ions into the adjacent extracellular microspace. Most of the observations made in the outer secretory enamel, i.e., selective B-type carbonate substitution, thin ribbons in crystal morphology (characteristic for OCP [65]), high content of acid phosphate, and shifts of PO_4^{3-} bands in FTIR spectra, are consistent with the initial precipitation of an OCP-like phase. On the other hand, X-ray or electron diffraction work on the youngest enamel failed to discern the presence of OCP reflections [20, 66]. High resolution electron microscopy demonstrated evidence for the vestige of OCP domains in the lattice of synthetic carbonatoapatites [67, 68], while no corresponding evidence has yet been attained for enamel crystallites. Altogether, there is still a lack of conclusive evidence for precipitation of OCP during enamel mineralization.

Future Prospects

The overall results support the contention that the nature of the biomineral(s) and its precipitation kinetics are determined primarily by the degree of supersaturation in the microenvironment surrounding the precipitating crystals, although the presence of inhibitors and promoters may substantially modify the situation. In order to gain a more complete picture about the regulatory mechanism of carbonatoapatite formation in biological systems, the experimental approaches and models (e.g., solubility model) mimic more closely the in vivo situation and the actual nature of the mineral. Consideration of carbonate and acid phosphate as constituents of the biomineral is certainly better than assigning as its stoichiometry that of hydroxyapatite, $(Ca)_5(PO_4)_3(OH)$. However, it should be noted that, as shown in Table 1, the enamel fluid (and other mineralizing fluids) contains significant concentrations of Mg^{2+} and Na^+. It was reported [69, 70] that parts of the magnesium and sodium in the enamel tissue are located in the crystal lattice. Currently, little information exists on what is the extent of incorporation of these ions into the crystal lattice or if the main portion is surface bound or, as suggested by others [71, 72], present in non-apatitic phases. Similarly, Young and Spooner [73] reported that the OH^- content of human enamel apatite is at least 20%–30% below the theoretical content for stoichiometric hydroxyapatite. However, no quantitative information about the OH^- contents in enamel crystals at various developing stages is available. A better understanding of the stability of enamel mineral and the driving force for the precipitation and growth of enamel crystals will be attained through further studies taking into account most of these putative lattice constituents. A model for enamel mineral of the type $Ca_{5-x}(Mg)_q(Na)_u(HPO_4)_v(CO_3)_w(PO_4)_{3-y}(OH,F)_{1-z}$ may better reflect the chemical nature of enamel mineral and may lead to a better understanding of its properties.

Acknowledgements. The authors are grateful to Drs. S. Shimoda and Y. Miake and Ms. C. Creighton for their valuable technical help. This work was supported by Grants DE08670, DE07623, DE07009, DE03187 from the National Institute of Dental Research and a grant from the Colgate-Palmolive Company.

References

1. Bonel G, Montel G (1964) Sur une nouvelle apatite carbonatee synthetique. C R Acad Sc Paris 258:923–926
2. Elliott JC (1964) The crystallographic structure of dental enamel and related apatites. Ph.D. Thesis, University of London
3. LeGeros RZ, Trautz OR, Klein E, LeGeros JP (1969) Two types of carbonate substitution in the apatite structure. Experientia 24:5–7
4. McClellan GH, Lehr JR (1969) Crystal-chemical investigations of natural apatites. Amer Min 54:1374–1391
5. Rey C, Collins B, Goehl T, Dickson RI, Glimcher MJ (1989) The carbonate environment in bone mineral: A resolution-enhanced Fourier transformed infrared spectroscopy study. Calcif Tissue Int 45:157–164
6. Rey C, Shimizu M, Collins B, Glimcher MJ (1990) Resolution-enhanced Fourier transformed infrared spectroscopy study of the environment of phosphate ions in the early deposits of a solid phase of calcium-phosphate in bone and enamel, and their evolution with age. Investigations in the mu_4 PO_4 domain. Calcif Tissue Int 46:384–394
7. Bachra BN, Fischer HRA (1969) The effect of some inhibitors on the nucleation and crystal growth of apatite. Calcif Tissue Res 3:348–357
8. Greenfield DJ, Termine JD, Eanes ED (1974) A chemical study of apatites prepared by hydrolysis of amorphous calcium phosphate in carbonate containing aqueous solutions. Calcif Tissue Res 14:131–138
9. Shimoda S, Aoba T, Moreno EC, Miake Y (1990) Effect of solution composition on morphological and structural features of carbonated calcium apatites. J Dent Res 69:1731–1740
10. LeGeros RZ, Trautz OR, LeGeros JP, Klein E, Shirra WP (1967) Apatite crystallites: Effects of carbonate on morphology. Science 155:1409–1411
11. Aoba T (1977) X-ray diffraction study on the amorphous and crystalline components in bone mineral (in Japanese). Jpn J Oral Biol 20:616–627
12. Nelson DGA, Featherstone JDB, Duncan JF, Cutress TW (1982) Paracrystalline disorder of biological and synthetic carbonate-substituted apatites. J Dent Res 61:1274–1281
13. Hallsworth AS, Weatherell JA, Robinson C (1973) Loss of carbonate during the first stages of enamel caries. Caries Res 7:345–348
14. LeGeros RZ, Tung MS (1983) Chemical stability of carbonate- and fluoride-containing apatites. Caries Res 17:419–429

15. Mayer I, Voegel JC, Bres EF, Frank RM (1988) The release of carbonate during the dissolution of synthetic apatite and dental enamel. J Crystal Growth 87:129–136
16. Brudevold F, Soremark R (1967) Chemistry of the mineral phase of enamel. In: Structural and chemical organization of teeth, vol. 1. Miles AEW (ed) Academic New York, pp 247–277
17. Weatherell JA, Robinson C, Hiller CR (1968) Distribution of carbonate in thin sections of dental enamel. Caries Res 2:1–9
18. Elliott JC, Holcomb DW, Young RA (1985) Infrared determination of the degree of substitution of hydroxyl by carbonate ions in human dental enamel. Calcif Tissue Int 37:372–375
19. Hiller CR, Robinson C, Weatherell JA (1975) Variations in the composition of developing rat incisor enamel. Calcif Tissue Res 18:1–12
20. Aoba T, Moreno EC (1990) Changes in the nature and composition of enamel mineral during porcine amelogenesis. Calcif Tissue Int 47:356–364
21. Howell DS, Pita JC, Marquez JF, Madruga JE (1968) Partition of calcium, phosphate, and protein in the fluid phase aspirated at calcifying sites in epiphyseal cartilage. J Clin Invest 47:1121–1132
22. Wuthier RE (1977) Electrolytes of isolated epiphyseal chondrocytes, matrix vesicles, and extracellular fluid. Calcif Tissue Res 23:125–133
23. Neuman WF (1969) The *milieu interieur* of bone: Claude Bernard revisited. Fed Proc 28:1846–1850
24. Aoba T, Moreno EC (1987) The enamel fluid in the early secretory stage of porcine amelogenesis: Chemical composition and saturation with respect to enamel mineral. Calcif Tissue Int 41:86–94
25. Nylen MU, Eanes ED, Omnell KA (1963) Crystal growth in rat enamel. J Cell Biol 18:109–123
26. Kerebel B, Daculsi G, Kerebel LM (1979) Ultrastructural studies of enamel crystallites. J Dent Res 58(B):844–850
27. Daculsi G, Menanteau J, Kerebel LM, Mitre D (1984) Length and shape of enamel crystals. Calcif Tissue Int 36:550–555
28. Miake Y, Aoba T, Moreno EC, Shimoda S, Prostak K, Suga S (1990) Ultrastructural studies on crystal growth of enameloid minerals in elasmobranch and teleost fishes. Calcif Tissue Int 48:204–217
29. Robinson C, Fuchs P, Deutsch D, Weatherell JA (1978) Four chemically distinct stages in developing enamel from bovine incisor teeth. Caries Res 12:1–11
30. Warshawsky H (1984) Ultrastructural studies on amelogenesis. In: The chemistry and biology of mineralized tissues. Butler WT (ed) Birmingham, Ala, pp 33–45
31. Suga S (1983) Comparative histology of progressive mineralization pattern of developing enamel. In: Suga S (ed) Mechanisms of tooth enamel formation. Quintessence, Tokyo, pp 167–203
32. Aoba T, Tanabe T, Moreno EC (1987) Function of amelogenins in porcine enamel mineralization during the secretory stage of amelogenesis. Adv Dent Res 1:252–260
33. Fejerskov O (1979) Human dentition and experimental animals. J Dent Res 58(B):725–731
34. Aoba T, Collins J, Moreno EC (1989) Possible function of matrix proteins in fluoride incorporation into enamel mineral during porcine amelogenesis. J Dent Res 68:1162–1168
35. Shimoda S, Aoba T, Mereno EC (1991) Changes in acid phosphate content in enamel mineral during porcine amelogenesis. J Dent Res 70:1516–1523

36. Aoba T, Moreno EC (1992) Changes in the solubility of enamel mineral at various stages of porcine amelogenesis. Calcif Tissue Int 50:266–272
37. Weatherell JA, Deutsch D, Robinson C, Hallsworth AS (1975) Fluoride concentrations in developing enamel. Nature 256:230–232
38. Speirs RL (1975) Fluoride incorporation into developing enamel of permanent teeth in the domestic pig. Arch oral Biol 20:877–883.
39. Crenshaw MA and Bawden JW (1981) Fluoride-binding by organic matrix from early and late developing bovine fetal enamel determined by lowrate dialysis. Arch Oral Biol 26:473–476
40. Lussi A, Fridell RA, Crenshaw MA, Bawden JW (1988) Absence of in vitro fluoride-binding by the organic matrix of developing bovine enamel. Arch Oral Biol 33:531–533
41. Shimizu M, Fukae M (1983) Enamel proteins. In: Suga S (ed) Mechanisms of tooth enamel formation. Quintessence, Tokyo, pp 125–141
42. Aoba T, Tanabe T, Moreno EC (1987) Proteins in the enamel fluid of immature porcine teeth. J Dent Res 66:1721–1726
43. Aoba T, Moreno EC, Tanabe T, Fukae M (1990) Effects of fluoride on matrix proteins and their properties in rat secretory enamel. J Dent Res 69:1248–1255
44. Varughese K, Moreno EC (1981) Crystal growth of calcium apatites in dilute solutions containing fluoride. Calcif Tissue Int 33:431–439
45. Phakey PP, Palamara JP, Conn A, Rachinger WA, Orams HJ (1989) Artifical remineralization of tooth enamel surfaces. In: Fearnhead RW (ed) Tooth enamel V. Florence, Yokohama, pp 504–508
46. Fowler BO, Moreno EC, Brown WE (1966) Infra-red spectra of hydroxyapatite, octacalcium phosphate and pyrolysed octacalcium phosphate. Arch Oral Biol 11:477–492
47. Sauer GR, Wuthier RE (1988) Fourier transform infrared characterization of mineral phases formed during induction of mineralization by collagenase-released matrix vesicles in vitro. J Biol Chem 263:13718–13718
48. Rey C, Beshah K, Griffin R, Glimcher MJ (1991) Structural studies of the mineral phase of calcifying cartilage. J Bone Min Res 5:515–525
49. Nancollas GH, Tomazic B (1974) Growth of calcium phosphate on hydroxyapatite crystals. Effect of supersaturation and ionic medium. J Phys Chem 78:2218–2225
50. Aoba T, Moreno EC (1984) Preparation of hydroxyapatite crystals and their behavior as seeds for crystal growth. J Dent Res 63:874–880
51. Aoba T, Yoshioka C, Yagi T (1984) Seeded crystal growth of calcium apatite and inhibitory activity of small organic molecules. In: Fearnhead RW, Suga S (eds) Tooth enamel IV. Elsevier, Amsterdam, pp 3–7
52. Aoba T, Fukae M, Tanabe T, Shimizu M, Moreno EC (1987) Selective adsorption of porcine amelogenins onto hydroxyapatite and their inhibitory activity on seeded crystal growth of hydroxyapatite. Calcif Tissue Int 41:281–289
53. Aoba T, Moreno EC (1989) Mechanism of amelogenic mineralisation in minipig secretory enamel. In: Fearnheard RW (ed) Tooth enamel V. Florence, Yokohama, pp 163–168
54. Aoba T, Moreno EC (1990) Structural relationship of amelogenin proteins to their regulatory function of enamel mineralization. In: Sikes CS, Wheeler AP (eds) Surface reactive peptides and polymers. American Chemical Society, pp 85–106
55. Patel PR, Brown WE (1975) Thermodynamic solubility product of human tooth enamel: Powdered sample. J Dent Res 54:728–736

56. Larsen MJ, Jensen SJ (1989) The hydroxyapatite solubility product of human dental enamel as a function of pH in the range 4.6–7.6 at 20°C. Arch Oral Biol 34:957–961
57. Moreno EC, Aoba T (1991) Comparative solubility of human dental enamel, dentin, and hydroxyapatite. Calcif Tissue Int 49:6–13
58. Vignoles M, Bonel G, Holcomb DW, Young RA (1988) Influence of preparation conditions on the composition of type B carbonated hydroxyapatite and on the localization of the carbonate ions. Calcif Tissue Int 43:33–40
59. Kay MI, Young RA, Posner AS (1964) Crystal structure of hydroxyapatite. Nature 204:1050–1052
60. LeGeros RZ, Suga S (1980) Crystallographic nature of fluoride in enameloids of fish. Calcif Tissue Int 32:169–174
61. Brown WE, Mathew M, Tung MS (1981) Crystalchemistry of octacalcium phosphate. Prog Cryst Growth Charact 4:59–87
62. Brown WE, Chow LC, Siew C, Gruninger S (1984) Acidic calcium phosphate precursors in formation of enamel mineral. In: Fearnhead RW, Suga S (eds) Tooth enamel IV. Elsevier, Amsterdam, pp 8–13
63. Chickerur NS, Tung MS, Brown WE (1980) A mechanism for incorporation of carbonate into apatite. Calcif Tissue Int 32:55–62
64. Moreno EC, Aoba T (1987) Calcium binding in enamel fluid and driving force for enamel mineralization in the secretory stage of amelogenesis. Adv Dent Res 1:245–251
65. Terpstra RA, Bennema P (1987) Crystal morphology of octacalcium phosphate: Theory and observation. J Crystal Growth 82:416–426
66. Landis WJ, Burke G, Neuringer JR, Paine MC, Nanci A, Bai P, Warshawsky H (1988) Earliest enamel deposits of the rat incisor examined by electron microscopy, electron diffraction, and electron probe microanalysis. Anat Rec 220:233–238
67. Nelson DGA, Wood GJ, Barry JC, Featherstone JDB (1986) The structure of (100) defects in carbonated apatite crystallites: A high resolution microscope study. Ultramicroscopy 19:253–266
68. Nelson DGA, Barry JC, Shields CP, Glena R, Featherstone JDB (1989) Crystal morphology, composition, and dissolution behavior of carbonated apatites prepared at controlled pH and temperature. J Colloid Interface Sci 130:467–479
69. LeGeros RZ (1984) Incorporation of magnesium in synthetic and in biological apatites. In: Fearnhead RW, Suga S (eds) Tooth enamel IV. Elsevier, Amsterdam, pp 32–36
70. Driessens FCM, Heijligers HJM, Borggreven JMPM, Woltgens JHM (1984) Variations in the mineral composition of human enamel on the level of cross-striations and striae of Retzius. Caries Res 18:237–241
71. Casciani FS, Etz ES, Newbury DE, Doty SB (1979) Raman microprobe studies of two mineralizing tissues: Enamel of the rat incisor and the embryonic chick tibia. Scan Electron Microsc II:383–390
72. Driessens FCM, Verbeeck RMH (1985) Dolomite as a possible magnesium-containing phase in human tooth enamel. Calcif Tissue Int 37:376–380
73. Young RA, Spooner S (1969) Neutron diffraction studies of human tooth enamel. Arch Oral Biol 15:47–63

Specific Physicochemical Properties of Fluoridated Hydroxyapatites with Incorporated Trace Elements

MASAYUKI OKAZAKI[1]

Summary. The crystallinity of fluoridated CO_3-apatites showed similar patterns of variation in the degree of fluoridation irrespective of carbonate content, although, on the whole, it was lower than that of fluoridated hydroxyapatites. In Mg-containing fluoridated hydroxyapatites, the patterns of crystallinity with the increase of fluoride content were unique in each series with different magnesium contents. In contrast to these unexpected crystallinity phenomena, the apparent solubility of fluoridated CO_3-apatites and Mg-containing fluoridated hydroxyapatites decreased monotonically with the increasing degree of fluoridation, and approached that of fluoridated hydroxyapatites at a high fluoride content. Fe-containing fluoridated hydroxyapatites were more poorly crystallized than Fe-free fluoridated hydroxyapatites. Despite this inhibited crystallinity behavior, the apparent solubility decreased more than that of Fe-free fluoridated hydroxyapatites at a low fluoride content.

Key words: Fluoridated apatites—CO_3^{2-}—Mg^{2+}—Fe^{2+}

Introduction

Previous investigators [1–3] reported that, from the physicochemical standpoint, incorporation of fluoride into the crystalline lattice of apatitic tooth mineral brings about greater stability. In general, crystallinity reflects the crystal's physicochemical strength, and, therefore, an increasing fluoride content may increase crystallinity, i.e., crystal strength. Additionally, the relationship of the crystallinity of fluoridated apatites to their solubility might be closely related to the susceptibility of caries to fluoridated enamel apatites. However, as caries-preventive agents, fluoride ions often act unpredictably.

F^- ions can be almost stoichiometrically taken up into the apatite crystals. Unless it is controlled, the environment of apatite formation in vivo and in vitro is unlikely to have a stoichiometric (Ca/P) value 1.67 of the hydroxyapatite.

[1] Osaka University Faculty of Dentistry, Suita, Osaka, 565 Japan

Hydroxyapatites prepared over a wide range (0.1–6.67) of calcium-phosphate feed molar ratios, $(Ca/P)_f$, are not strongly affected by these ratios [4]. However, F^- uptake of fluoridated apatites synthesized at low $(Ca/P)_f$, i.e., below 1.67, is strongly inhibited [5]. Furthermore, the composition of the mineral phase is not exactly the same as that of the standard hydroxyapatite. There are various other constituents present. In the presence of trace elements such as $CO_3{}^{2-}$, Mg^{2+}, Fe^{2+}, etc., the formation of fluoridated hydroxyapatites may become more complex.

Computer Graphics of Fluoridated Hydroxyapatites

Information on the conformation of hydroxyapatite in relation to its crystallinity, solubility, and/or mechanical strength is useful for understanding the mechanism and prevention of dental caries. After the expenditure of a great deal of time and effort, a ball-and-stick model of hydroxyapatite using plastic spheres has been successfully devised. Recently, computer graphics have made remarkable progress: while a large computer was needed a few years ago to draw a molecular model, this can now be easily achieved with only a personal computer and a CRT [6]. The benefit of computer graphics is in making it possible to view a model from any direction, and to instantly change its elements and their sizes and positions. Therefore, hydroxyapatite substituted by a trace element can be viewed immediately. The structure of hydroxyapatite has been analyzed by using X-ray diffraction and neutron diffraction analysis [7].

Figure 1.A shows a computer graphic of fluoridated hydroxyapatite along the *c*-axis, in which F^- ions are partially substituted into OH^- positions. Two columns of Ca^{2+} ions are visible in the unit cell. If computer graphics were drawn simply using Pauling's ionic radius [8], P^{5+}(0.33Å) and H^+(≒0Å) ions would be lost. Therefore, each ionic radius was modified as follows: O^{2-}(1.40Å → 1.20Å), P^{5+}(0.33Å → 0.70Å), H^+(≒0Å → 0.50Å), F^-(1.36Å → 1.10Å). The ionic radius of Ca^{2+}(0.99Å) was not changed. By rotating this view through −90° along the X-axis of the CRT, Fig. 1.B was obtained.

The crystallographic structure of hydroxyapatite had been difficult to analyze, however, because of the shifted position of OH^- ions, which is located about 0.3Å above or below the stable position of the crystal [1]. Now the instability of the structure can be easily understood using the computer graphics of fluoridated hydroxyapatite, in which F^- ion is located just on the mirror plane at $Z = 1/4$. This computer graphics display demonstrates that fluoride ions serve to stabilize the hydroxyapatite crystal and prevent dental caries. The computer graphics are helpful in understanding the physicochemical properties of fluoridated apatites with trace elements.

CO_3-Containing Fluoridated Hydroxyapatites

A few investigators have touched briefly upon the subject of fluoridated CO_3-apatites. In most studies, however, both the synthetic and tooth apatites examined were carbonate-free fluoridated hydroxyapatite. It is well known that

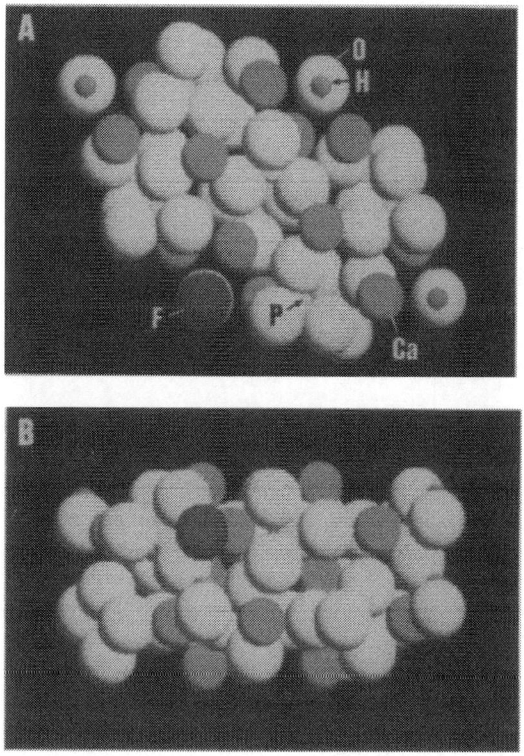

Fig. 1. Computer graphics of a fluoridated hydroxyapatite structure **A** projected on the X-Y plane and **B** rotated through −90° around the X-axis

the inorganic phases of tooth enamel are mainly carbonate-containing apatites and that carbonate in the apatites may weaken the chemical bonds in the crystals [9]. Thus, it is important to examine the interaction between the caries-preventive action of fluoride and the caries-promoting action of carbonate in apatite crystals.

We have reported that crystallinity of synthetic fluoridated hydroxyapatites showed no simple increase with the degree of fluoridation, while solubility showed a simple decrease [10]. We suspect that the fluoride content may also have a strong effect of the physicochemical properties of carbonate-containing apatites.

Fluoride contents of the fluoridated CO_3-apatites, FCO_3Ap_I and FCO_3Ap_{II}, when synthesized with carbonate-phosphate feed molar ratios, $(CO_3/P)_f = 0.5$ and 2.0, increased with the fluoride concentration in the solution [11]. Fluoride could be taken up almost stoichiometrically. Neither calcium nor phosphate contents changed significantly with changes in fluoride content. Notably, carbonate content was not affected by fluoride content. The X-ray diffraction patterns of all samples were characteristic of calcium apatites. Samples showed a lower degree of crystallization with increase of the carbonate content. No extraneous peaks due to CaF_2 were found, indicating that the fluoride content of samples did not exceed 2 mmol/g, which is equal to that of francolite (carbonate fluorapatite).

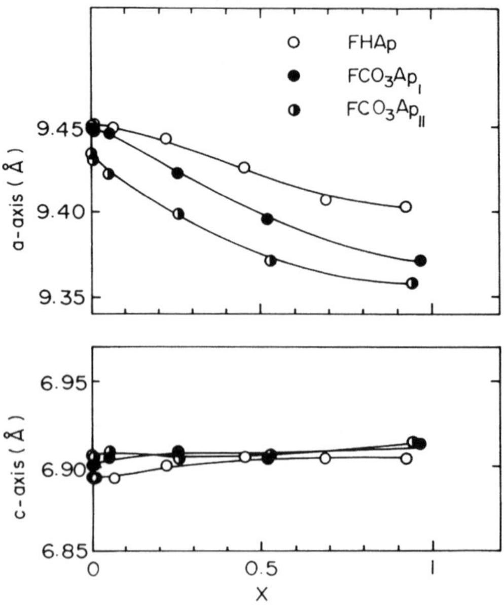

Fig. 2. *A*- and *c*-axis dimensions of fluoridated CO_3apatites as a function of the degree of fluoridation, X

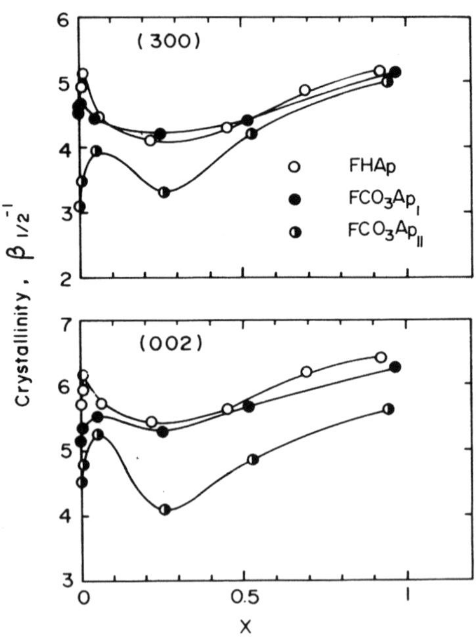

Fig. 3. Crystallinity of fluoridated CO_3-apatites shown as the inverse of the half-value breadths of the (300) and (002) reflections

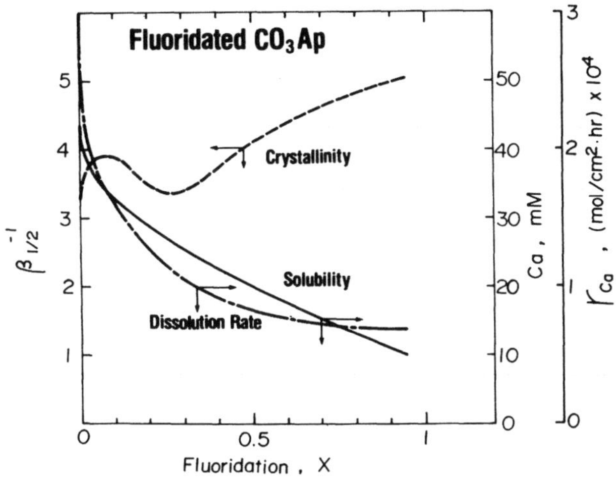

Fig. 4. Crystallinity, solubility, and dissolution rate behaviors of fluoridated CO_3-apatites

Figure 2 shows the *a*- and *c*-axis dimensions of each sample as a function of fluoride content. With an increase in the degree of fluoridation, X, the *a*-axis dimensions decreased, whereas the *c*-axis dimensions changed very little. This result agrees with the observations on fluoridated HAp by Moreno et al. [3]. These and the results of chemical analysis indicate that fluoridated CO_3-apatites were produced. In addition, the *a*-axis dimensions of samples with high carbonate content, FCO_3Ap_{II}, decreased more than those with low carbonate content, FCO_3Ap_I.

Figure 3 shows the crystallinity of fluoridated CO_3-apatites as the inverse of the half-value breadths of the (300) and (002) reflections. The crystallinity of each fluoridated CO_3-apatites initially increased slightly, then decreased with the degree of fluoridation, and finally increased greatly when the fluoride content approached that of francolite, analogous to that of the fluoridated hydroxyapatites, FHAp. It should be noted that the crystallinity of each fluoridated CO_3-apatite decreased at lower fluoride contents.

We were able to observe an interesting phenomenon in the magnified infrared absorption spectra in the region of 1000–800 cm^{-1} [12]. The peak due to CO_3^{2-} ions at 875 cm^{-1} shifted to a lower frequency with the degree of fluoridation. This means that the conformation of CO_3^{2-} ions in the apatite crystal may be greatly affected by the fluoride content, even though the carbonate content was not affected by the degree of fluoridation.

On the other hand, the solubility and dissolution rate of fluoridated CO_3-apatites decreased dramatically at low fluoride content levels and continued to gradually decrease monotonically. Thus, no analog to solubility behavior was found in the complex phenomena of fluctuating crystallinity (Fig. 4). The disorder and/or random distribution of F^- ions in the crystals may possibly weaken the crystallinity and/or inhibit the crystal growth at relatively low fluoride content levels.

Mg-Containing Fluoridated Hydroxyapatites

Recently, clinical attention is being focused on magnesium because of its close relation to the metabolism and physiology of the human body [13]. Bone and teeth also contain certain amounts of magnesium, which may play an important role in the initial formation of apatites and have a significant effect on their physicochemical properties. For example, magnesium may control the rate of apatite formation in the early stages of mineralization in a matrix vesicle (MV) [14]. Furthermore, it can be speculated that magnesium may associated with fluoride ions which promote the apatite formation.

The chemical analysis of Mg-containing hydroxyapatites indicated that with an increasing magnesium-calcium feed molar ratio, $(Mg/Ca)_f$, the calcium content decreased and the magnesium content increased [15]. However, the phosphate content showed no significant change below $(Mg/Ca)_f = 0.5$. The $(Ca + Mg)/P$ molar ratio of the sample of $(Mg/Ca)_f = 1$ was close to that of tricalcium phosphate, 1.5, while the other samples showed apatitic values close to the apatitic figure of 1.67 [15].

Mg-containing hydroxyapatites became poorly crystallized at higher ratios and showed an amorphous structure at $(Mg/Ca)_f = 1$. However, none of the samples gave a tricalcium phosphate pattern, even though the chemical composition of the sample of $(Mg/Ca)_f = 1$ had a similar value 1.45 of $(Ca + Mg)/P$ to tricalcium phosphate 1.5.

Although the change in *a*-axis dimensions was less distinct because of the rapid appearance of amorphous patterns, *c*-axis dimensions decreased with the increase of magnesium content. These and the results of chemical analysis suggest that Mg^{2+} ions may substitute for Ca^{2+} ions. We can then suggest that Mg^{2+} ions can substitute for calcium positions to a limited extent, and that a small amount of magnesium may decrease the crystallinity of hydroxyapatite and make it soluble.

Both the calcium and phosphate contents of Mg-containing fluoridated hydroxyapatite synthesized with the magnesium-calcium feed molar ratio, $(Mg/Ca)_f = 0.01$ ($Mg\text{-}FHAp_I$) and 0.05 ($Mg\text{-}FHAp_{II}$) [16], did not vary significantly with an increasing fluoride content (X: the degree of fluoridation). Magnesium content, on the other hand, increased with an increase in the degree of fluoridation, X (Fig. 5), indicating that magnesium uptake may be promoted by fluoride. This result agrees with the observation of McCann and Bullock [17].

The *a*-axis dimensions of both $Mg\text{-}FHAp_I$ and $Mg\text{-}FHAp_{II}$ decreased with an increased degree of fluoridation, in the same way as that of Mg-free fluoridated hydroxyapatites, FHAp, suggesting a substitution of F^- ions into the apatite crystals. The *a*- and *c*-axis dimensions of both $Mg\text{-}FHAp_I$ and $Mg\text{-}FHAp_{II}$ were lower than those of FHAp. The *c*-axis dimension, in particular, decreased with the increase of magnesium content over the whole range of fluoride content, suggesting a substitution of magnesium ions into the apatite crystals.

Figure 6 shows the crystallinity of the 2 series of Mg-containing fluoridated apatites as the inverse of the half-value breadths calculated from the (300) and

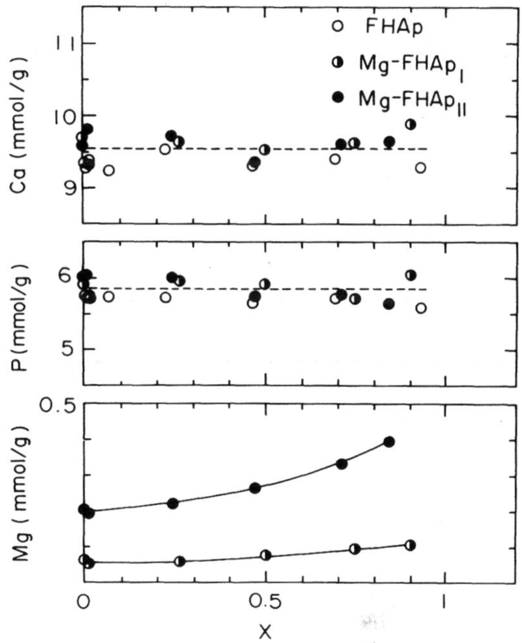

Fig. 5. Chemical compositions of Mg-containing fluoridated hydroxyapatites as a function of the degree of fluoridation, X

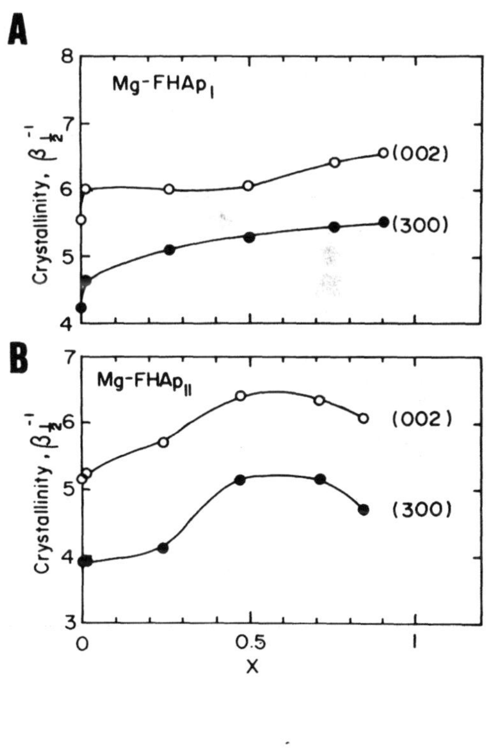

Fig. 6. Crystallinity of Mg-containing fluoridated hydroxyapatites, **A** Mg-FHAp$_I$ and **B** Mg-FHAp$_{II}$, shown as the inverse of the half-value breadths of the (300) and (002) reflections. (See Fig. 3 for Mg-free fluoridated hydroxyapatites, FHAp)

(002) reflections, which are taken as representations of the crystallinity along the *a*- and *c*-axis. In fluoridated hydroxyapatites, FHAp, the crystallinity increased initially, then decreased, and finally increased (Fig. 3). In the series with lower Mg contents, Mg-$FHAp_I$, the minimum observed at low fluoride content of Mg-free fluoridated hydroxyapatites was obliterated (Fig. 6.A); and, in the series with higher Mg-content, Mg-$FHAp_{II}$, crystallinity rose steadily to a maximum at about one-half of the maximum fluoride content (Fig. 6.B), although crystallinity of the F-free Mg-containing hydroxyapatites decreased monotonically with the increasing Mg content [15].

The apparent solubility of Mg-containing fluoridated hydroxyapatites $FHAp_I$ and $FHAp_{II}$ decreased monotonically with the increase of fluoride content in the same way as that of Mg-free fluoridated hydroxyapatites, FHAp. These 3 apatite series did not show any significant differences when approaching a high fluoride content. In the solubility behaviors, the action of magnesium is behind the strong stability action of fluoride.

Fe-Containing Fluoridated Hydroxyapatites

Iron, an indispensable element for human life, forming the core of hemoglobin molecules, Hb(Fe^{2+}), in red blood cells [13] for example, is also a trace element in tooth minerals. In East Asia, the tooth black "ohaguro," that was traditionally applied to teeth, contained iron, which could have contributed to caries prevention [18]. The effects on caries-preventive action of cations substituting Ca^{2+} sites, and of fluoride substituting OH^- sites have been studied [19]. Some fish species have a large quantity of iron and fluoride in their teeth [20]. Thus, it is important to clarify the efficiency of fluoride in combination with iron in caries prevention.

Hydroxyapatites synthesized by using $FeCl_2 \cdot nH_2O$ rapidly became poorly crystallized with the increase of ferrous ions in the solution [21] (Fig. 7). In addition to apatite, at a higher concentration of ferrous ions, a different phosphate was formed, (Fig. 7.C, D). This was identified as $Ca_3Fe_4(PO_4)_4$-$(OH)_6 \cdot 3H_2O$, in which iron was oxidized to the ferric state. Its X-ray patterns clearly differed from those of the iron sources, $FeCl_2 \cdot nH_2O$ and $FeCl_3 \cdot 6H_2O$.

Ferrous (Fe^{2+}) ions seemed to be taken up by apatite crystals. The contraction of *a*- and *c*-axis dimensions of Fe-containing apatites with the increase of iron content is suggested, although the differences found are too small to be conclusive. However, the peak shifts of the expanded (300) and (002) reflections shown in Fig. 7.E support this suggestion.

The X-ray diffraction patterns of Fe-containing fluoridated apatites synthesized with the iron-calcium feed molar ratio $(Fe^{2+}/Ca)_f = 0.07$ [22] were similar to those of calcium apatites, but the degree of crystallization was greatly reduced in comparison with that of Fe-free fluoridated hydroxyapatites as previously reported [10, 11]. Calcium contents of samples were lower than those of Fe-free fluoridated hydroxyapatites, but the total molar contents of calcium and iron were similar to those of Fe-free fluoridated hydroxyapatites.

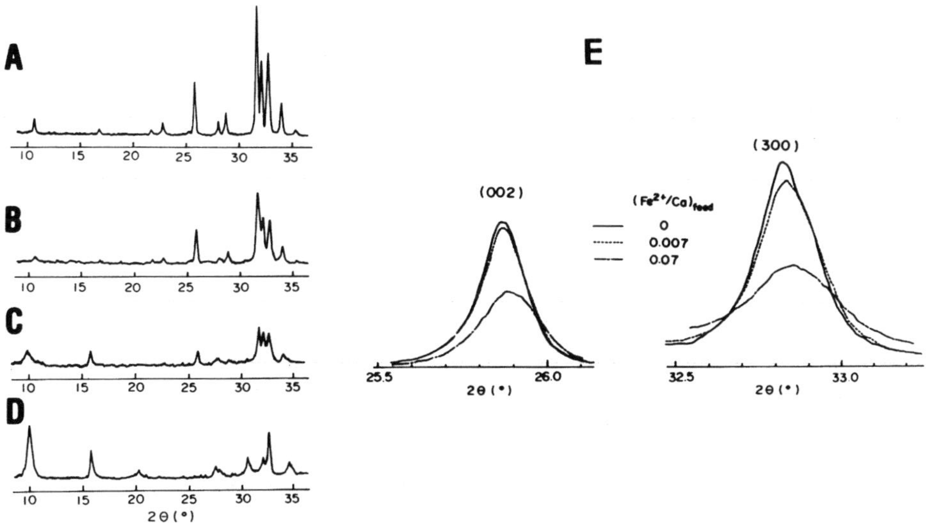

Fig. 7. X-ray diffraction patterns of Fe-containing hydroxyapatites (*A*–*D*) and their expanded (300) and (002) reflections (*E*). $(Fe^{2+}/Ca)_f$ = 0 (**A**); 0.07 (**B**); 0.37 (**C**); 0.73 (**D**)

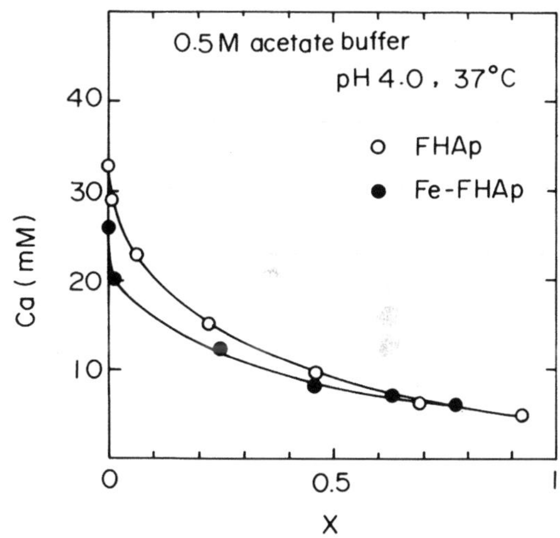

Fig. 8. Apparent solubility of Fe-containing fluoridated apatites shown as the dissolved calcium concentration in the solution

Calcium, iron, and phosphate contents were almost independent of the degree of fluoridation, X.

A-axis dimensions of Fe-containing fluoridated apatites decreased with the degree of fluoridation, in the same way as those of fluoridated hydroxyapatites [10, 11, 16]. The contraction was enhanced in both *a*- and *c*-axis dimensions, suggesting that Fe^{2+} ions may substitute into the fluoridated apatite crystals.

Figure 8 shows the apparent solubility of Fe-containing fluoridated apatites and those of Fe-free fluoridated apatites previously synthesized [10, 11]. At first, the solubility decreased greatly and then gradually with the degree of fluoridation, whereas the crystallinity showed a complex behavior. The solubility of F-free Fe-containing hydroxyapatite was significantly lower than that of F- and Fe-free hydroxyapatite in spite of the decrease in crystallinity. Such decrease in calcium concentration by the dissolution-preventive action of iron is clear at very low fluoride concentrations, even if the decrease in the relative calcium content of apatites that accompanies the substitution on Fe^{2+} ions is considered. However, at a high fluoride content, both curves seem to converge, and the additional effect of Fe^{2+} ions seems to disappear.

A number of investigators have studied the combined effects of various compounds with fluoride. Often, however, the efficiency of these compounds is masked by the dramatic caries-preventive action of fluoride. In this study, the effectiveness of iron in caries prevention has been clearly demonstrated. It is interesting that this dissolution-preventing action of iron was enhanced at very low concentrations of fluoride, irrespective of crystallinity.

References

1. Kay MI, Young RA, Posner AS (1964) Crystal structure of hydroxyapatite. Nature 204:1050–1052
2. Miles AEW (1967) Structural and chemical organization of teeth, vol. II. Academic, New York
3. Moreno EC, Kresak M, Zahradnik RT (1977) Physicochemical aspects of fluoride-apatite systems relevant to the study of dental caries. Caries Res 11 (Suppl 1): 142–171
4. Okazaki M, Aoba T, Doi Y, Takahashi J, Moriwaki Y (1979) The effect of the calcium-phosphate feed molar ratios on apatite formation. J Osaka Univ Dent Sch 19:87–93
5. Okazaki M, Takahashi J, Kimura H (1985) F^- uptake inhibition by excess during fluoridated apatite formation. Caries Res 19:342–347
6. Okazaki M, Sato M (1990) Computer graphics of hydroxyapatite and β-tricalcium phosphate. Biomaterials 11:573–578
7. Sudarsanan K, Young RA (1965) Significant precision in crystal structure details: Holly springs hydroxyapatite. Acta Cryst B25:1534–1543
8. Pauling L (1960) The nature of the chemical bond, 3rd edn. Cornell University Press, Ithaca
9. LeGeros RZ (1967) Apatite crystallites: Effects of carbonate on morphology. Science 155:1409–1411
10. Okazaki M, Aoba T, Doi Y, Takahashi J, Moriwaki Y (1981) Solubility and crystallinity in relation to fluoride content of fluoridated hydroxyapatites. J Dent Res 60:845–849
11. Okazaki M, Takahashi J, Kimura H, Aoba T (1982) Crystallinity, solubility, and dissolution rate behavior of fluoridated CO_3apatites. J Biomed Mater Res 16:851–860

12. Okazaki M (1983) F^{-}-CO_3^{2-} interaction in IR spectra of fluoridated CO_3-apatites. Calcif Tissue Int 35:78–81
13. Stryer L (1981) Biochemistry. W.H. Freeman, San Francisco
14. Anderson HC (1967) Electron microscopic studies of induced cartilage development and calcification. J Cell Biol 23:81–101
15. Okazaki M, Takahashi J, Kimura H (1986) Unstable behavior of magnesium-containing hydroxyapatites. Caries Res 20:324–331
16. Okazaki M (1987) Mg^{2+}-F^{-} interaction during hydroxyapatite formation. Magnesium 6:296–301
17. McCann HG, Bullock FA (1957) The effect of fluoride ingestion on the composition and solubility of mineralized tissues of the rat. J Dent Res 36:391–398
18. Yamaga R, Nakayama M (1975) Ohaguro (tooth black) and its prophylactic effects. Kokusai Shika J 1:401–414
19. Francis MD, Gray JA, Griebstein WJ (1968) The formation and influence of surface phases on calcium phosphate solid. Arch Oral Biol 3:83–120
20. Suga S (1984) The role of fluoride and iron in mineralization of fish enameloid. In: Fearnhead RW, Suga S (eds) Tooth enamel IV. Elsevier, Amsterdam, pp 472–477
21. Okazaki M, Takahashi J, Kimura H (1985) Iron uptake of hydroxyapatite. J Osaka Univ Dent Sch 25:17–24
22. Okazaki M, Takahashi J, Kimura H (1986) Crystallinity and solubility behavior of iron-containing fluoridated hydroxyapatites. J Biomed Mater Res 20:879–886

The Effect of Fluoride on Enamel De- and Remineralization

JACOB M. (BOB) TEN CATE[1]

Summary. Fermentation of carbohydrates in the dental plaque leads to the formation of acids, causing the pH to drop and enamel to demineralize. In between such acid attacks, the condition of supersaturation of the oral fluids to hydroxyapatite results in a precipitation of calcium phosphate mineral at the sites depleted of mineral. Recent observations have shown that the processes of demineralization and remineralization are both affected significantly by fluoride, even at very low concentrations, in the oral fluids. Such fluoride levels may be achieved by topical fluoride applications, drinking water fluoridation, or the use of fluoridated dentifrices. The rationale for the inhibition of demineralization and the enhancement of remineralization is thermodynamic: when fluoride is present, mineral deposition, in the form of a fluoridated hydroxyapatite, is enhanced due to its lower solubility product. Inhibition of demineralization is the result of the reprecipitation of a fluoride-rich mineral during enamel dissolution. Moreover, the phase thus formed is less susceptible to future acid attacks.

This paper was presented during a symposium held at the Nippon Dental University, Tokyo, Japan, October 13, 1990.

Key words: Enamel—Demineralization—Remineralization—Fluoride

Introduction

For many years, caries was considered to be an inevitable disease. Once caries had begun to develop at a particular site in the dentition, it progressed rapidly, forming a cavity which needed dental treatment in the form of a restoration. However, in the 1960s evidence was presented that caries need not necessarily progress to deep cavities but, instead, that a caries defect may also be remineralized or be arrested by the influx of salivary constituents. Koulourides et al. [1] immersed enamel lesions in saliva in a laboratory experiment and showed that the uptake of calcium and phosphate from saliva results in a rehardening of the enamel lesions. This process was described as remineralization.

[1] Department of Cariology and Endodontology, Academic Centre for Dentistry, Amsterdam (ACTA), The Netherlands

At the same time, in 1966, Backer Dirks [2] presented data from a clinical study which indicated that remineralization has an impact on the rate at which caries develops in the mouth as well. In brief, these findings revealed that about 50% of the white spot lesions detected on the buccal surfaces of the first molars of 8-year-old children disappeared during the ensuing (7) years due to the remineralizing capacity of saliva, particularly in geographical areas where the water had been fluoridated.

Mechanistic Studies of Enamel Remineralization

With the clinical findings of enamel remineralization, the possible implications of this phenomenon in caries prevention were rapidly understood: if saliva was able to counteract caries attacks in a natural way, this process could possibly lead to an easy and effective method of caries inhibition. Therefore, many studies were initiated with the purpose of understanding the mechanisms of enamel remineralization, with the ultimate aim of using this knowledge to develop caries preventive treatments and regimens. Some of these studies will be discussed here, and the conclusions derived from these studies will be given.

In the 1950s and 1960s, it had become possible to mimic the process of forming white spot caries lesions in the laboratory. Several model systems had been developed to create enamel lesions with the histological characteristics of caries as it develops in the oral cavity. These included systems comprised of gelatin [3] and other gels [4] to simulate the dental plaque, and of organic acids which were similar to the acid formed in dental plaque during the fermentation of carbohydrates. Thus, different systems were available for forming white spot caries lesions in the laboratory. This obviously had the advantage that the lesions could be produced in large numbers and were easily reproducible, a necessary condition when these lesions had to be remineralized.

In studies on the mechanism of enamel lesion remineralization, artificially produced white spot lesions were generally used as the substrate material. A typical set-up for such studies involved the use of lesions and solutions mimicking saliva in its inorganic composition. Of course the organic components were considered equally important but, in the initial stages, emphasis was given to the mineral components of the saliva system. Thus, the solutions used resembled saliva in its degree of supersaturation. Nevertheless, they differed from saliva in the calcium-to-phosphate ratio. Where saliva typically is rich in phosphate and relatively low in calcium, the solutions used for remineralization studies had a calcium-to-phosphate ratio stoichiometric to hydroxyapatite (i.e., Ca/P = 1.66) [5, 6]. When solutions with a calcium content of 2.0 mMol/L, a phosphate content of 1.2 mMol/L, and a pH of 7 were used to remineralize lesions, a deposition of mineral occurred, as concluded from the depletion of ions from the remineralizing solutions [5–7] (Fig. 1).

Judging from the calcium-to-phosphorous ratio of the depositing material, this mineral was either apatite or a similar calcium phosphate mineral. Additional studies were conducted to look into the effect of adding fluoride to saliva-

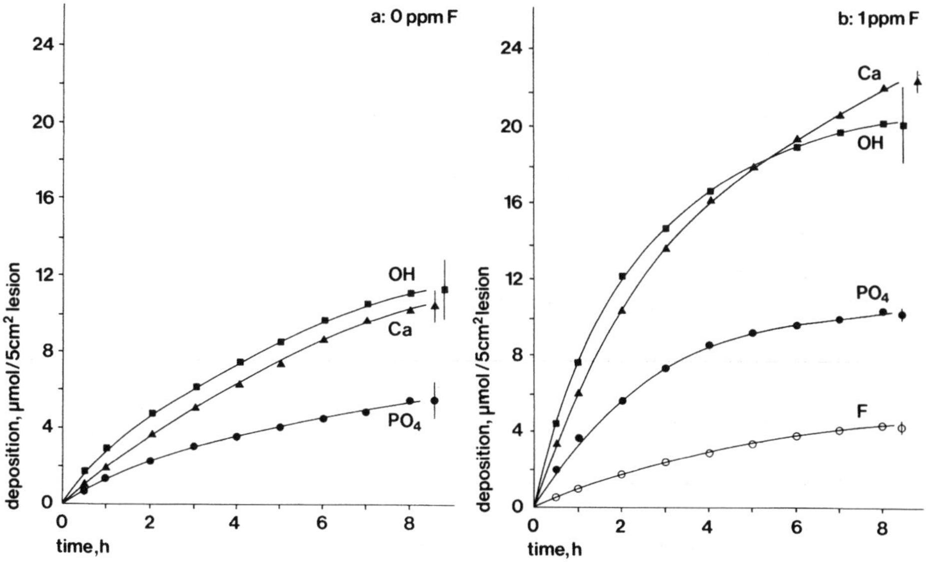

Fig. 1. Amounts of *Ca*, *P*, *OH*, and *F* deposited during 8-h immersion of 5 cm^2 of enamel lesion in 12.5 ml solution, initially containing 2.0 mMol/L $CaCl_2$, 1.2 mMol/L KH_2PO_4, and 150 mMol/L KCl at pH = 7.0 and a constant fluoride concentration of either **a** 0 (no added fluoride) or **b** 1 ppm. (With permission from [7])

like solutions, and it was found that even very low fluoride concentrations in solution would significantly speed up the apatite deposition [7–9]. The physico-chemical rationale for this was that, in the presence of fluoride, not only hydroxyapatite but also a fluoridated hydroxyapatite can form, the latter having a lower solubility and thus a higher thermodynamic driving force for precipitation.

Apart from such physico-chemical descriptive and analytical chemical studies, the lesions remineralised in vitro were analyzed using histological, histomorphometric, and tracer techniques. For instance, using the scanning electron microscope, it was found that artificial lesions were depleted of mineral, as could be seen from a thinning of crystallites in the most demineralized part of the lesions (the "lesion body"). When such lesions were subjected to remineralizing solution, it was found that, in particular, the thickness of the crystallites increased. During remineralization, new crystals were also formed in the specimens [10].

These findings indicate that not only did remineralisation occur primarily by the growth of existing crystals but also that new crystals could be formed by secondary nucleation onto existing crystals or other remnants which had remained in the lesion after demineralization.

The finding that lesion remineralization resulted in an increase of the crystallinity and of the crystallites' sizes was later confirmed in a study by Silverstone [11]. He found that, in particular, the crystallites in the surface layer and in the

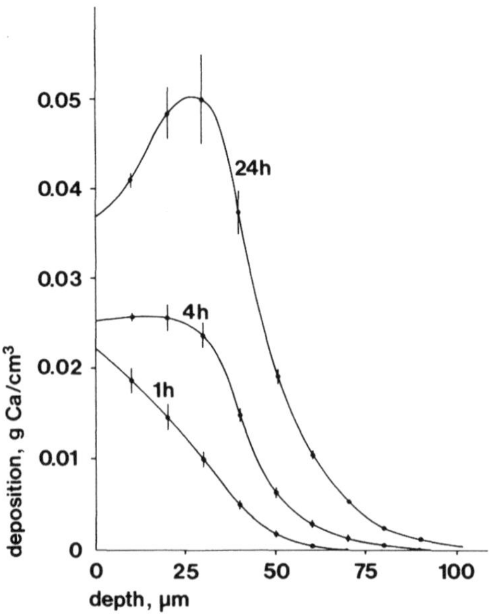

Fig. 2. Location of deposition within a lesion after 1, 4, and 24 h immersion in solutions initially containing 1.5 mMol/L $CaCl_2$, 0.9 mMol/L KH_2PO_4, and 150 mMol/L KCl at pH = 7.0. (With permission from [12])

dark zone (the region just beyond the lesion body) were increased in thickness, demonstrating that these were the regions where mineral deposition occurred during the remineralization process. Independent of these electron microscopic studies, experiments were performed with radioactive (labelled) calcium in the remineralizing solution in order to identify the sites in the lesion where deposition occurred. These studies, in which thin layers were removed by grinding off specimens which were then analyzed for radioactivity, revealed that, after an initial lag period, the radioactive deposition concentrated in the lesion body [12] (Fig. 2).

The general conclusions drawn from the numerous studies on mechanistic aspects are that remineralization is a process that can be simulated in the laboratory, that remineralization from solutions with a supersaturation similar to saliva occurs in the form of the deposition of an apatite-like mineral, and that this precipitation results in a growth of the partly demineralized crystals inside the lesion as well as the formation of some new crystals. This latter finding serves to explain why remineralized lesions never become as densely mineralized as the original, sound enamel.

This phenomenon, which could be called "scar tissue formation", is particularly evident when fluoride is provided. The rationale is that the accelerated deposition of a fluoride containing mineral occurs less specifically inside the lesion and may even occur in the narrow pores of the surface layer. This could result in lesions which are completely remineralized superficially, leaving the underlying tissue still demineralized [10, 13] (Fig. 3). This phenomenon has also been observed clinically. Backer Dirks' early observation [2] revealed that, in areas with drinking water fluoridation, many of the white spot lesions had

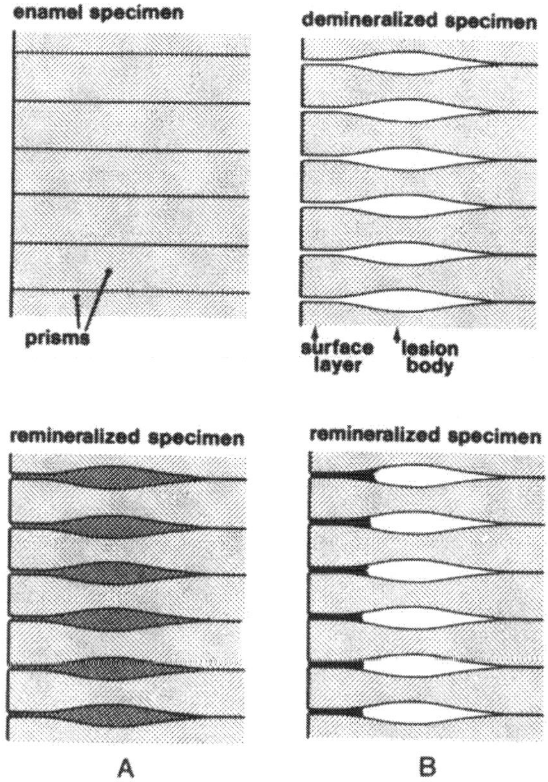

Fig. 3. A Schematic representation of the formation and (partial) remineralization of a caries lesion. When viewed in cross section, lesions form by the dissolution of mineral from the interprismatic regions, preferentially from sites underneath the anatomical surface. Remineralization may occur by a redeposition of mineral at the sites attacked. **B** Alternatively, the pores in the surface layer of the lesions may be filled up by mineral, causing a surface layer blocking and resulting in arrested (consolidated) lesions. (Adapted from [13])

a glossy appearance which was attributed to surface remineralization [14]. Thus, although the lesions still appeared white, it was concluded that such arrested lesions would not progress into cavities.

Another general finding of these studies was that very low fluoride concentrations in solution were able to significantly alter the rate of remineralization, and this again is of significant practical importance. When fluoride-containing dentifrices are used, this results in an elevation of the fluoride levels in the oral fluids. Obviously, the fluoride levels are increased substantially immediately after brushing, but even long after the actual brushing, a two-fold increase in fluoride baseline levels is still observed.

A similar situation exists when fluoridated drinking water is used. The physiological fluoride levels are also increased by about two-fold in those cases. The observation that remineralization is enhanced in areas with drinking water fluoridation [2, 14] serves to demonstrate that fluoride dentifrices, which yield a similar increase in fluoride levels in the oral fluids, improve the rate at which remineralization may occur. Therefore, the conclusion is justified that, for the prevention of dental caries, it is important to give remineralization an optimal chance. This is achieved by using fluoride dentifrices to increase fluoride levels in the oral fluid.

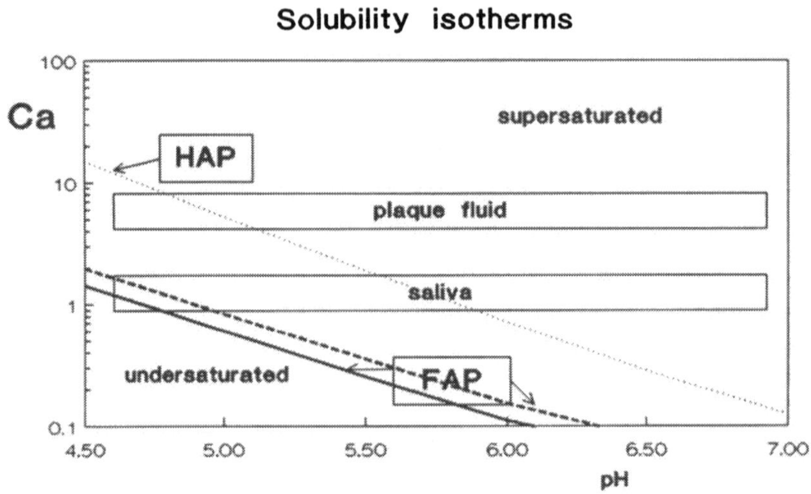

Fig. 4. Solubility isotherms for hydroxyapatite (*HAP*) and fluorapatite (*FAP*) at fluoride concentrations in solution of 1 and 10 ppm, respectively. For FAP, the *full line* represents the saturation calcium concentrations at 10 ppm F/L and the *dashed line* is the corresponding calcium levels at 1 ppm F/L. Calcium levels in saliva and plaque fluid, indicative of the degree of supersaturation with respect to the two minerals, are represented schematically

The Effect of Fluoride on Enamel Demineralization

Having established the effect of low fluoride levels in the ambient fluids on mineral deposition, a study was initiated on the effect of sub ppm levels of fluoride on enamel demineralization. Thermodynamically, the rationale for there being an effect of fluoride on enamel demineralization is very similar to its effects on remineralization. If enamel is attacked by weak acids in the presence of fluoride, a situation could arise where hydroxyapatite will dissolve but fluoridated apatite could still precipitate (Fig. 4). This would mean that while enamel dissolves, the calcium and phosphate ions could reprecipitate with the fluoride supplied from solution [15, 16]. Studies aimed to determine the effects of low fluoride levels on enamel demineralization have indicated that fluoride levels in the range of 0.05–5 ppm were very effective in inhibiting demineralization [17–19]. It was concluded that the fluoride concentration is as important as the pH in determining the rate at which a lesion develops, to such an extent that lesion formation may even be completely inhibited. The fluoride concentration at which inhibition occurs is dependent not only on the pH of the plaque, but also on the calcium and phosphate levels of the plaque. Together, they determine the driving force for the precipitation of fluoridated apatite at the sites under caries attack. Combined laboratory and clinical studies indicate that an effect of ambient fluoride on enamel demineralization is feasible at fluoride concentrations prevailing in the oral fluids.

In addition to the noted effects of fluoride on the rate of lesion formation, it was also observed that fluoride had an effect on the histology of the forming lesions. When fluoride was added to solution, lesions would form with a very thick surface layer, whereas in fluoride-depleted or fluoride-free solutions, a caries defect would form with an eroded aspect. Taking this observation one step further, it could be argued that many of the white spot lesions observed on extracted teeth are lesions which had already been remineralized to a significant extent, because such lesions would typically have quite pronounced surface layers which were rich in fluoride. Thus, the question of whether a surface layer is a phenomenon associated with caries lesion formation or whether it is the result of a remineralization process is yet unsolved, and both hypotheses still stand at present.

pH-Cycling

While the mechanistic studies on enamel demineralization and enamel remineralization were conducted, techniques were also developed which enabled the monitoring of the pH fluctuations in dental plaque in the course of time, in particular after eating fermentable carbohydrates. From the results it was concluded that caries never occurs as a continuous process, but rather during short periods of acid attack, lasting about 30 to 45 min. During major parts of the day, the dentition is bathing in saliva of near neutral pH from which mineral may be redeposited. This continuous fluctuation of the pH during the day initiated a new type of research which is now commonly referred to as "pH-cycling" [20]. In such experiments, the enamel is subjected to conditions of de- and remineralization mimicking the conditions of super- and subsaturation occurring in the mouth during the day. During the past decade, this method has been used in many laboratories [21–23] and has been advanced in different ways to such an extent that actual "Stephan curves" (giving the change in pH after acid production in the plaque) can be simulated in the laboratory [24]. These studies are of value for answering fundamental questions regarding the mineral uptake and loss by either enamel lesions or by sound enamel under different physiological conditions, as well as for studying the effects of caries preventive agents necessary to formulate optimal treatments.

Using the former approach, it was shown that low fluoride levels, present during the demineralization phase or during remineralization, were effective in shifting the net balance of demineralization and remineralization to the positive site (i.e., mineral accretion). Typically, using low fluoride levels during pH-cycling, the initial effect would be on the inhibition of demineralization during the caries attack cycles. Gradually, the fluoride also affects the mineral uptake as the inhibition of caries breakdown limits the number of sites available for regrowth of material. In other words, the amount of turnover of mineral is decreased after the addition of fluoride (Fig. 5).

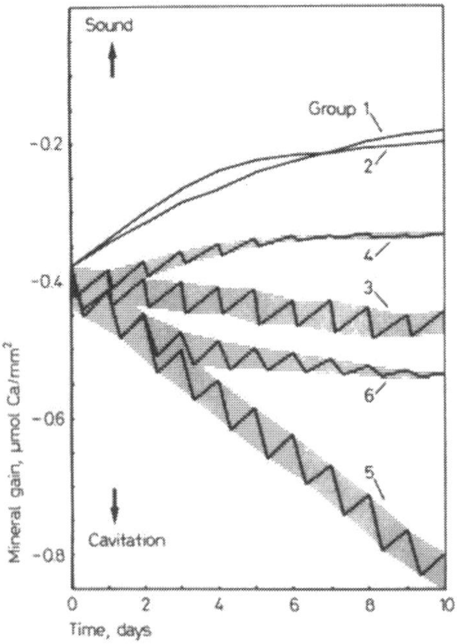

Fig. 5. Cumulative plot of calcium loss and gain for different cycling schemes with or within fluoride added at 2 ppm concentration to the remineralization cycles. In groups 1 and 2 the specimens were subjected to remineralization only. In groups 3 and 4 there was a daily 3-hour demineralization cycle and a 21-hour remineralization cycle. In groups 5 and 6, these time periods comprised 6 and 18 h, respectively. Before the start of the experiment, incipient caries lesions had been formed in the specimen, as can be deduced from the value of $-0.376\,\mu mol\,Ca/mm^2$ mineral content at $t = 0$. The figures shows that the cycling scheme used and the presence of fluoride are parameters determining whether lesion progression or arrest occur. (With permission from [20])

Intraoral Studies

Even though the recently developed laboratory methods take into account many of the intraoral parameters, the caries mechanisms can never be completely and definitively unraveled from in vitro studies. Therefore, intraoral models were developed to study the caries formation in the mouth and to study enamel remineralization. Such studies involve the use of volunteers who usually wear dentures onto which enamel specimens are attached. This can be done in several ways [25]. One application is the use of partial dentures in which enamel specimens are placed in the buccal flange in the molar region. Specimens can be pretreated in the laboratory, e.g., to create incipient caries lesions, before they are placed in the mouth. Also, the specimens may be covered with gauze which would accumulate plaque formation [26].

The aims to conduct in situ studies have been threefold. First, intraoral models have been used to study the initiation and progression of caries condit-

Fig. 6. Results obtained for a group of 18 participants in an intraoral study to assess the degree of lesion repair when a placebo paste, a fluoride paste, and a fluoride + pyrophosphate (anti-tartar) paste were compared. The percentage of lesion repair (positive values) is given. In some individuals, lesions were found to further demineralize, as indicated by the negative repair values. Panellist initials are given at the x-axis, as are the group averages

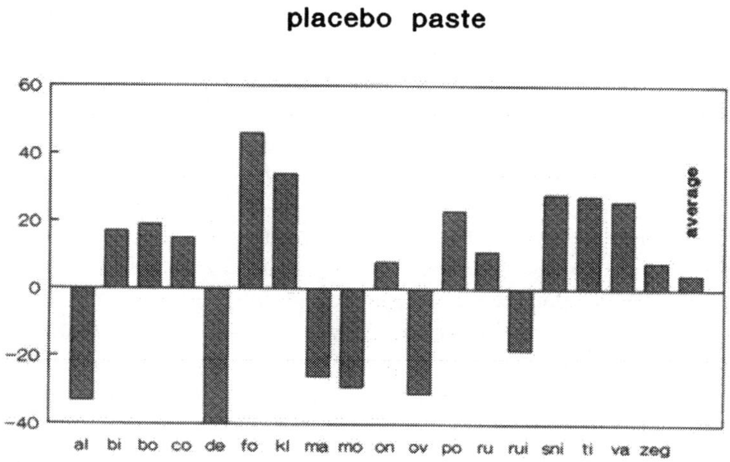
placebo paste
60
40
20
0
-20
-40
average
al bi bo co de fo kl ma mo on ov po ru rui sni ti va zeg

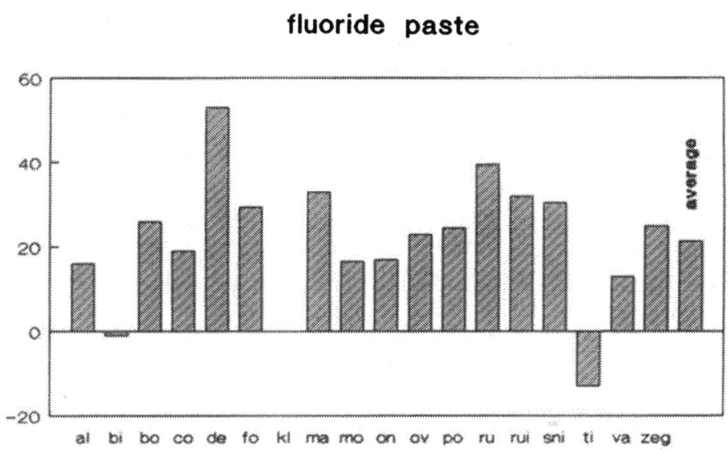
fluoride paste
60
40
20
0
-20
average
al bi bo co de fo kl ma mo on ov po ru rui sni ti va zeg

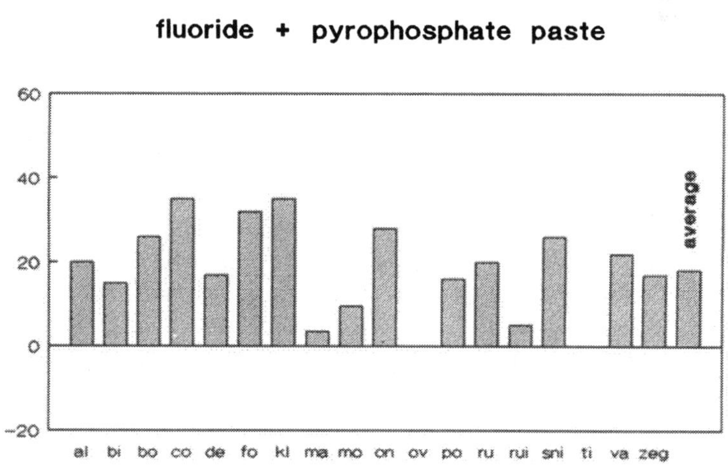
fluoride + pyrophosphate paste
60
40
20
0
-20
average
al bi bo co de fo kl ma mo on ov po ru rui sni ti va zeg

ions involving extraoral sucrose rinses with or without a caries preventive agent as additive. Second, studies have been conducted to determine the amount of fluoride uptake as a result of a given preventive scheme. Third, intraoral studies have been aimed at determining the remineralization of preformed lesions, in particular also studying the effect of caries preventive agents in their normal usage on this process. Unlike in the laboratory where conditions can be monitored continuously and can be made optimal for enamel remineralization in the mouth, the pH and the calcium and phosphate levels of the fluids in which the specimens are bathing vary continuously with time within each individual as well as between individuals. Therefore, it is not surprising that either caries progression or reversal may be found among individuals with the same experimental protocol. The same is even reported for individuals in subsequent trials [27]. In the mouth, a complicating factor is the presence of many organic constituents in saliva and plaque fluid which affect the process of mineral dissolution and precipitation. In saliva, many inhibitors have been identified which interact with the formation of calcium phosphates by lowering the driving force for precipitation due to binding with calcium. Other components bind strongly to mineral surfaces, resulting in inhibition of crystal growth. Therefore, in addition to the rates of mineral deposition and dissolution being determined by the mineral components of saliva and plaque fluid, an overriding effect may be the content and composition of saliva in terms of specifically binding proteins. In spite of their limitations, intraoral models give an insight into the magnitude and rate at which caries progression and lesion remineralization occur in individuals, and consequently these data enable researchers to fine-tune their laboratory models. A typical study of this type concerns a comparison of a fluoride and a placebo toothpaste with an anti-tartar paste. The aim of this study was whether the addition of anti-tartar ingredients resulted in an impaired anti-caries efficacy of the fluoride product. Eighteen panellists were asked to have enamel specimens with incipient caries lesions placed in their partial dentures during each of 3 7-week experiments. In the first leg of the study, a placebo paste was supplied to all participants. Then, in the next two legs of the study, the people were randomly assigned to one of the experimental groups which brushed their dentures with either a fluoride dentifrice or a fluoride anti-tartar dentifrice. At the end of each 7-week period, the specimens were removed and subjected to microradiographic analysis to determine the mineral content patterns. A comparison of the lesions before and after the in situ period indicated whether the experimental conditions had led to an enhancement of the caries defect or to a repair of the lesion. The result of this study is depicted in Fig. 6. From the data a number of conclusions can be drawn: irrespective of the type of dentifrice used there was a large inter-individual difference in intraoral response. For a placebo paste further demineralization was observed in some volunteers while remineralization occurred in others. With both types of fluoride dentifrices (with or without pyrophosphate as anti-tartar active) lesion repair was observed in all but one of the specimens, illustrating the fluoride enhancing effect on enamel remineralization.

In conclusion, intraoral studies show that the process of remineralization may contribute significantly to a slowing down, arrest, or even reversal of caries progression. Fluoride causes a shift in the demineralization-remineralization equilibrium in the mouth. Such effects can be observed even at the very low fluoride levels in the oral fluids such as those produced by fluoride supplementation schemes used in caries prevention.

References

1. Koulourides T, Feagin F, Pigman W (1965) Remineralisation of dental enamel by saliva in vitro. Ann NY Acad Sci 131:751–757
2. Backer Dirks O (1966) Posteruptive changes in dental enamel. J Dent Res 45: 503–511
3. Silverstone LM (1973) Structure of carious enamel, including the early lesion. In: Melcker AH, Zarb GA (eds) Dental enamel, development, structure and caries. Munksgaard, Copenhagen, pp 100–160
4. Gray JA (1966) Kinetics of enamel dissolution during formation of incipient caries-like lesions. Arch Oral Biol 11:397–422
5. Feagin FF (1971) Calcium, phosphate and fluoride depositions on enamel surfaces. Calcif Tissue Res 8:154–164
6. Feagin FF, Gonzalez M, Jeansonne BG (1972) Kinetic reactions of calcium phosphate and fluoride ions at the enamel surface-solution interface. Calcif Tissue Res 10:113–127
7. Cate JM ten, Arends J (1977) Remineralisation of artificial enamel lesions in vitro. Caries Res 11:277–286
8. Feagin FF, Jeansonne BG, (1973) Effective fluoride concentrations to promote apatite mineralisation at the enamel surface. Ala J Med Sci 10:107–114
9. Koulourides T, Cueto H, Pigman W, (1961) Rehardening of softened enamel surfaces on human teeth by solutions of calcium phosphates. Nature 189:226
10. Cate JM ten, Jongebloed WL, Arends J, (1981) Remineralisation of artificial enamel lesions in vitro, IV. Influence of fluorides and diphosphonates on short and long term remineralisation. Caries Res 15:60–69
11. Silverstone LM (1983) Remineralisation and enamel caries: New concepts. Dental Update 10:4
12. Cate JM ten, Arends J (1980) Remineralisation of artificial enamel lesions in vitro, III. A study of the deposition mechanism. Caries Res 14:351–358
13. Cate JM ten (1983) A model for enamel lesion remineralisation. In: Leach SA, Edgar WM (eds) Demineralisation and remineralisation of the teeth. IRL, Oxford, pp 129–144
14. Groeneveld A (1985) Longitudinal study of prevalence of enamel lesions in a fluoridated and non-fluoridated area. Community Dent Oral Epidemiol 13:159–163
15. Larsen MJ, Jensen SJ (1985) An X-ray diffraction and solubility study of equilibration of human enamel-powder suspensions in fluoride-containing buffer. Arch Oral Biol 30:471–475
16. Duff EJ (1976) Fluoride incorporation into powdered human enamel under conditions of fluctuating pH. Caries Res 10:234–240
17. Cate JM ten, Duijsters PPE (1983) Influence of fluoride in solution on tooth demineralisation. I Chemical data. Caries Res 17:193–199

18. Cate JM ten, Duijsters PPE (1983) Influence of fluoride in solution on tooth demineralisation. II. Microradiographic data. Caries Res 17:513–519
19. Margolis HC, Moreno EC, Murphy BJ (1986) Effect of low levels of fluoride in solution on enamel demineralization in vitro. J Dent Res 65:23–29
20. Cate JM ten, Duijsters PPE (1982) Alternating demineralisation and remineralisation of artificial enamel lesions. Caries Res 16:201–210
21. Featherstone JDB, O'Reilly MM, Shariati M, Brugler S (1986) Enhancement of remineralisation in vitro and in vivo. In: Leach, SA (ed) Factors relating to demineralisation and remineralisation of the teeth. IRL, Oxford, pp 23–35
22. Gerrard WA, Winter PJ (1986) Evaluation of toothpastes by their ability to assist rehardening of enamel in vitro. Caries Res 20:209–217
23. White DJ (1987) Reactivity of fluoride dentifrices with artificial caries. Caries Res 21:126–140
24. Almqvist H, Lagerlöf F, Angmar-Mansson B (1990) Automatic pH-cycling caries model applied on root hard tissue. Caries Res 24:1–5
25. Wefel JS (1990) Effects of fluoride on caries development and progression using intra-oral models. J Dent Res 69:626–633
26. Koulourides T, Phantumvanit P, Munksgaard EC, Housch T (1974) An intra-oral model used for studies of fluoride-incorporation in enamel. J Oral Path 3:185–195
27. Mellberg JR, Petrou ID, Deutchman M, Grote N (1988) The effects of 1% pyrophosphate and 0.02% sodium fluoride on artificial caries lesions in vivo. J Dent Res 67:1461–1465

Calcification in Octocorals

NORIMITSU WATABE[1] and RONI J. KINGSLEY[2]

Summary. The octocorals are colonial cnidarians, many of which contain calcareous spicules in the mesoglea and, in some species, in the axis as well. The octocorals are excellent organisms for probing the mechanisms of biomineralization and demineralization, and the processes of calcification of spicules have been investigated extensively in recent years, especially in *Leptogorgia virgulata*. In this review, octocoral spicule formation is discussed relative to (1) the characteristics of scleroblasts, the spicule-forming cells, (2) calcium transport by the colonies, (3) rate of calcification, (4) role of carbonic anhydrase, (5) hormonal control of calcification, and (6) organic matrices and their possible functions in spicule formation.

Key words: Octocorals — Spicules — Organic matrix — Ca-transport — Carbonic anhydrase

Introduction

The octocorals are colonial cnidarians which are comprised of animals of 6 orders: the Stolonifera (e.g., stolon corals, organ-pipe corals), Telestacea, Alcynoacea (soft corals), Coenothecalia (blue corals), Gorgonacea (e.g., sea whips, sea fans, and red corals), and Pennatulacea (e.g., sea pens and sea pansy). Many octocorals contain calcareous spicules in the mesoglea and also in the axis in some species. The spicule is generally spindle-shaped in form with various modifications characterizing different taxonomic categories [1]. The octocorals are excellent organisms for probing the mechanisms of biomineralization and demineralization. Thus, the processes of calcification of spicules have been investigated extensively in recent years, especially in the sea whip *Leptogorgia virgulata* (Fig. 1).

[1] Electron Microscopy Center, The University of South Carolina, Columbia, SC 29208, USA
[2] Department of Biology, University of Richmond, Richmond, VA 23173, USA

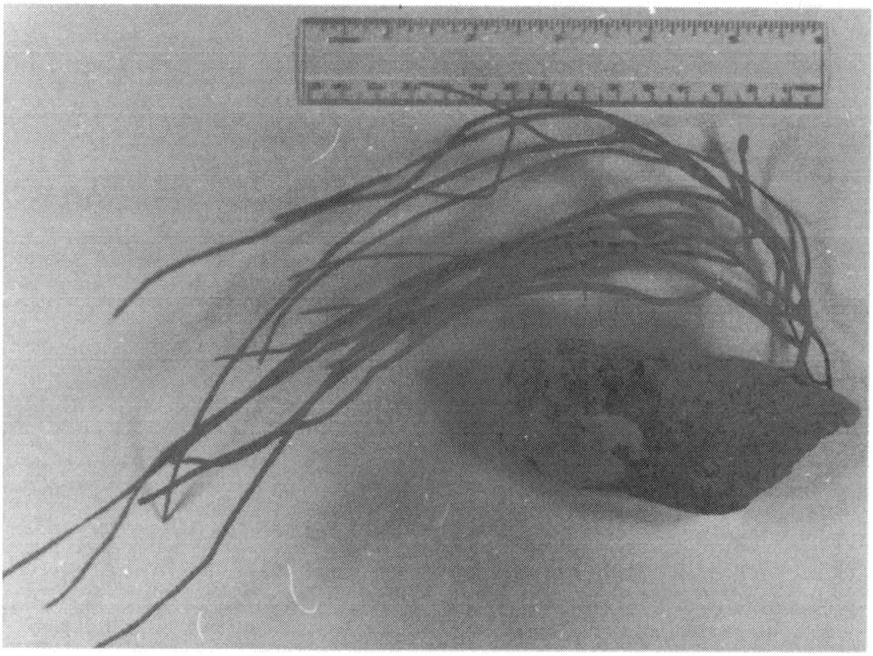

Fig. 1. Colonies of the gorgonian *Leptogorgia virgulata* attached to an oyster shell

Many octocoral spicules are composed of high magnesium calcite with 6–12 wt% magnesium carbonate [2]. The content of magnesium carbonate in all branch regions of the gorgonian is constant within a species [3]. The mature spicule of *Leptogorgia virgulata* is usually up to a few hundred μm long, and exhibits a more or less symmetrical spindle shape with many branches or small wart-like surface ornamentations [4] (Fig. 2). The spicule of the sea pansy *Renilla reniformis* is approximately 200 μm long and 25 μm wide, trilobed in cross sections, and characterized by the presence of grooves running parallel or obliquely to the morphological long axis of the spicule [5]. In both *Leptogorgia* and *Renilla*, the spicule is not of a single crystal but a polycrystalline aggregate of minute crystallites [4, 5]. In *Leptogorgia*, the surface ornamentation frequently shows aggregates of several rhombohedral calcite (Kingsley and Watabe, unpublished work). In cross sections, the crystallites are arranged in concentric layers, and organic matrix materials are present between the layers as well as within the crystals [4–7] (Fig. 3).

The Scleroblasts and Processes of Spicule Formation

The spicules are formed by the scleroblasts. The scleroblasts are the mesogleal cells derived from interstitial cells that have migrated inward from the epidermis [1]. In *Leptogorgia virgulata*, the preliminary phase of spicule formation

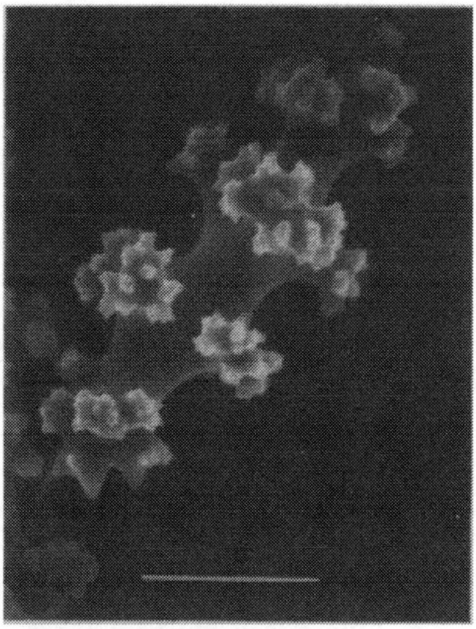

Fig. 2. Isolated mature spicule of *Leptogorgia virgulata*. SEM. *Bar* = 60 μm

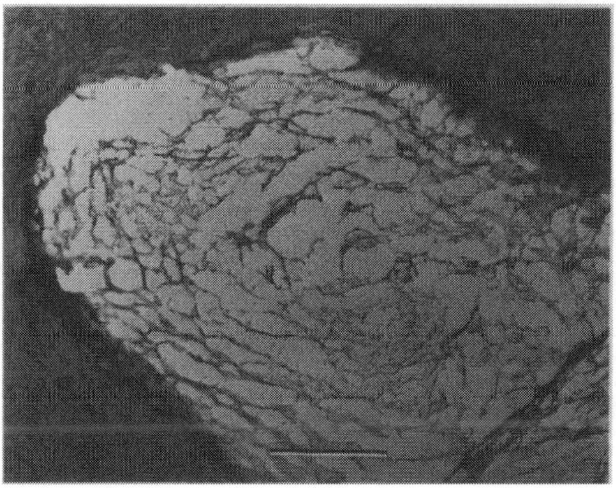

Fig. 3. Insoluble organic matrix. TEM of spicule section stained and decalcified with uranyl acetate. The matrix framework is more or less concentrically arranged. Perforations represent areas where calcite crystallites were present prior to decalcification. *Leptogorgia virgulata*. *Bar* = 1 μm

is the aggregation of the scleroblasts. Each cell contains a large nucleus, prominent Golgi apparatus, rough endoplasmic reticulum, free ribosomes, mitochondria, electron-dense bodies, vacuoles, and vesicles. A deposition of fibrous organic matrix within a vacuole, probably derived from the Golgi apparatus, is the first stage of spicule formation. The first crystals are laid down in close association with fibrous matrix materials in the vacuoles (Fig. 4).

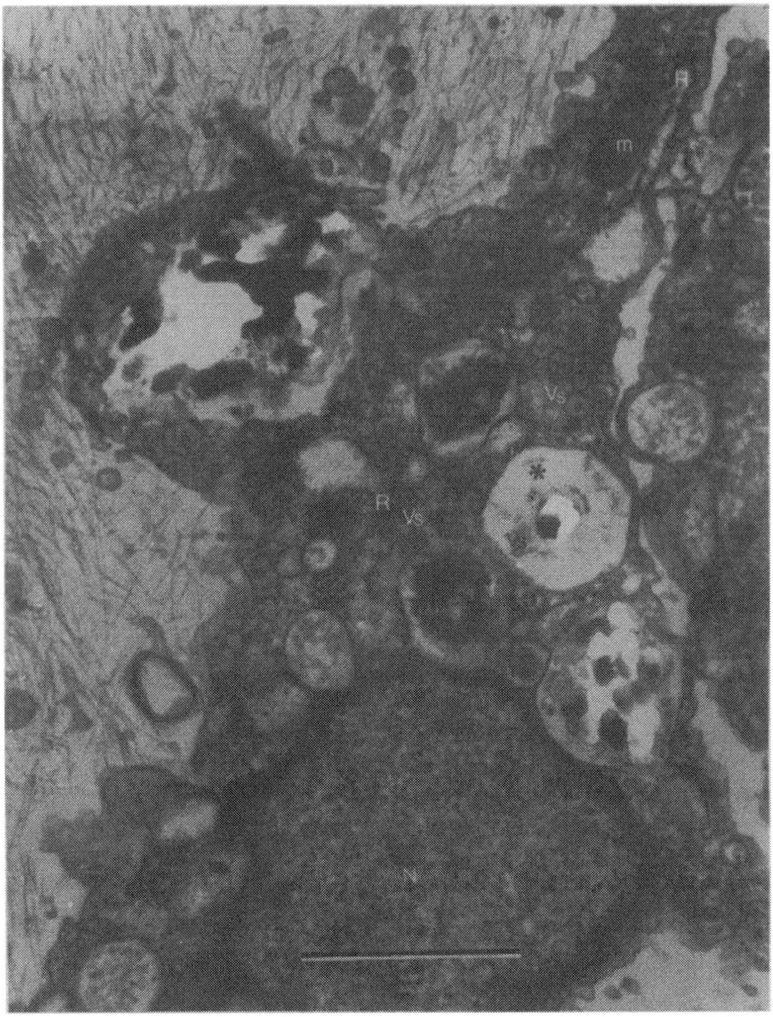

Fig. 4. TEM of early deposition of calcite crystals on organic matrix (*) within vacuoles of the scleroblast. *Leptogorgia virgulata*. Vs, Vesicles; *R*, rough endoplasmic reticulum; *N*, nucleus; m, mitochondrion. *Bar* = 1 μm. (With permission from [4])

The matrix and crystals continue to be deposited until the vacuole is almost completely occupied by the spicule. At or slightly before this stage, the scleroblasts dissociate from the aggregate and disperse. As the spicule grows, the vacuole enlarges and the nucleus becomes deformed and displaced toward the periphery of the scleroblast. The cytoplasm occupies a narrow space between the plasma membrane and the enlarged vacuole membrane. Wart-like structures become prominent at this stage. One of the most significant phenomena is the transition of the spicule from the intracellular to the extracellular entity.

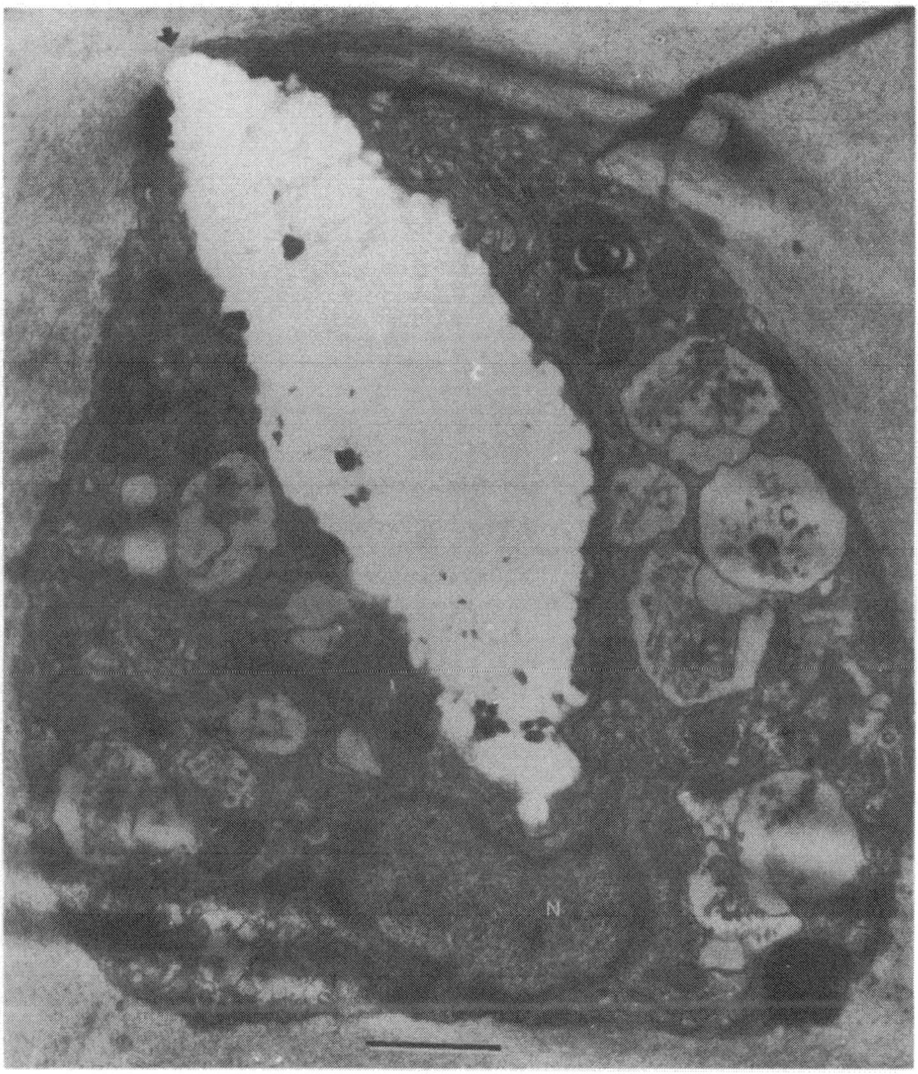

Fig. 5. TEM of a scleroblast of *Leptogorgia virgulata* showing maturing spicule (the *oblong space in the center*, the spicule crystals having been dislodged during sectioning) and a breach in the plasma membrane (*arrow*). *N*, Nucleus. *Bar* = 1 µm. (With permission from [4])

This is accomplished by a fusion of the spicule vacuole membrane with the plasma membrane of the scleroblast, followed by a breach formation in the scleroblast (Fig. 5). This causes the spicule to extend into the extracellular space. This extrusion has also been confirmed by cell culture of the scleroblast [8] (Fig. 6). Further growth of the spicule in the extracellular environment has

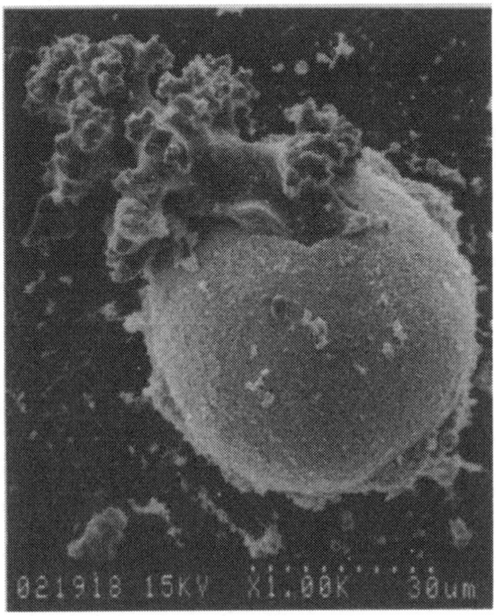

Fig. 6. Cultured scleroblast of *Leptogorgia virgulata* with mature spicule extruding from the cell. SEM. (With permission from [8])

not been confirmed, although it is possible that the growth occurs by the secretory activities of the original and/or secondary scleroblast [9]. In the gorgonian *Pseudoplexaura flagellosa*, scleroblast vesicles calcify and the crystals are released to the mesoglea [6]. Several of those primordial crystals integrate into a single mature spicule through the action of several secondary scleroblasts. How those primordial crystals assemble into a regularly shaped spicule is not clear. Further investigation is needed to clarify the additional growth of spicules in the extracellular environment.

Processes of formation of axial spicules have not been investigated in detail. In the primary polyp of the sea pen *Ptilosarcus gurneyi*, crystal development at an early stage of axis formation has been reported to occur in stylet cells in the endoderm [10]. However, in another sea pen *Veretillum cynomorium* the initial crystal-like deposits are not preformed within the epithelium surrounding the axis. They are presumed to be formed by precipitation from inorganic ions secreted into the general compartment of the axis [11]. Initial mineral deposits give rise to calcitic nodules which enlarge by the radial growth of several lobes. These nodules fuse and form the core of the axis (Fig. 7).

Calcium Transport for the Spicule Formation

Calcium transport for the spicule formation in *Leptogorgia virgulata* has been investigated by Kingsley and Watabe by ^{45}Ca pulse-chase experiments [12], ^{45}Ca light- and EM-autoradiography [13], and ATPases localization at the EM level [14]. The calcium movement from the seawater into the spicule vacuole and the axis occurs in several steps. First, calcium ions enter the outer epithe-

Fig. 7. Longitudinal fracture surface of a fully mineralized part of the axis of the sea pen *Veretillum cynomorium* after etching. Mineral columns (co) are radiating out from the axis core (c). SEM. *Bar* = 60 µm. (With permission from [11])

lium through diffusion (Fig. 8①). From the epithelium, these ions are actively transported out into the mesoglea (Fig. 8②) and also back to the seawater by the activity of Ca-ATPase localized at the epithelial plasma membranes. The exit of ions to the seawater is mediated largely through the polyp outer epithelium rather than the other outer epithelium. High activities of Ca-ATPase are found on the inner plasma membranes of the microvilli and lateral and basal plasma membranes of the polyp epithelium. From the mesoglea, some calcium ions diffuse into the scleroblasts. However, the majority of the calcium moves into the axis, which is the initial and primary site of calcium accumulation, through the axial epithelium (Fig. 8③④). The microvilli of the attachment cells (democytes) penetrate deep into the axis and inner membranes of those microvilli show very strong activities of Ca-ATPase. Approximately 30% of the calcium ions remain in the axis, probably being deposited as amorphous calcareous granules. The other 70% leave the axis and return to the mesoglea

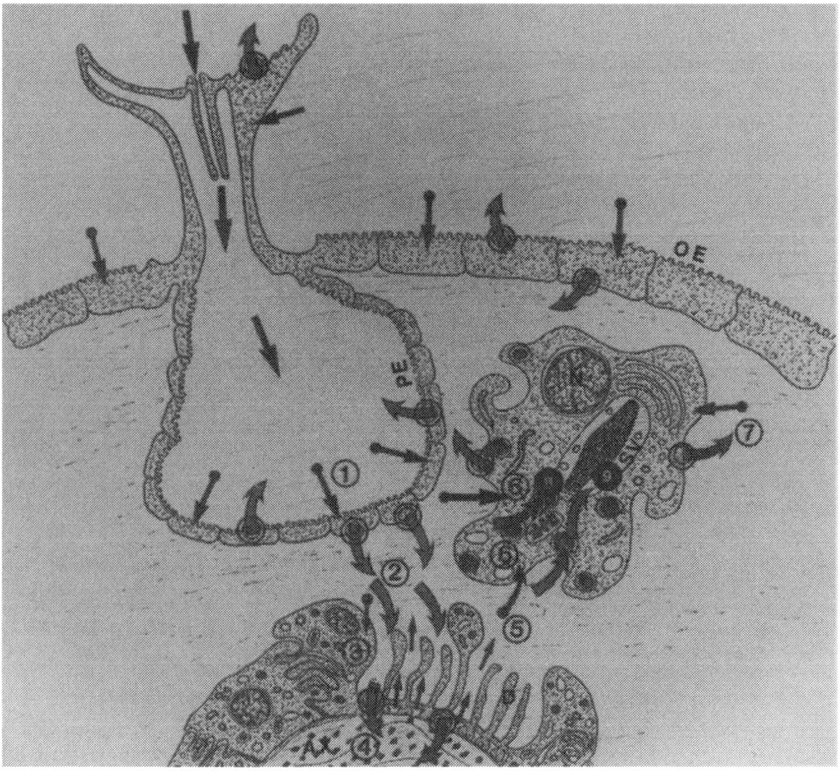

Fig. 8. A diagram depicting Ca^{2+} pathways from the environment into the spicule forming vacuole of the scleroblast of *Leptogorgia virgulata*. *Dark arrows* with *small dark circle* indicate diffusion. *Pale arrows* with *double circles* indicate active transport. *Other arrows* indicate directions of Ca^{2+} movement. *OE*, Outer epithelium; *PE*, polyp epithelium; *SV*, spicule vacuole; *N*, nucleus of the scleroblast; *AX*, axis; *D*, democyte

(Fig. 8⑤), and diffuse into the scleroblasts (Fig. 8⑥). They are further transported to the spicule-forming vacuole of the scleroblast via electron-dense bodies (Fig. 8⑧) to form the spicule (Fig. 8⑨). Portions of the Ca ions diffused into the scleroblasts are pumped back to the mesoglea (Fig. 8⑦). The plasma membrane of the scleroblasts and the limiting membrane of the electron-dense bodies also show ATPase activities. The autoradiographic study indicated that ^{45}Ca accumulates in the axis from 1 to 15 min after the 15-min incubation of the colony [13]. After 15 min, the label moves out of the axis. By 60 min, much of the label is in the spicules, reaching its maximum level at 120 min.

The calcium ions which leave the axis but do not enter the scleroblasts are released into the surrounding seawater. This backflow of calcium from the axis is a process also involving the active transport. This is different from the other gorgonian *Eunicella papillosa* and hermatypic corals in which the backflow is considered to be a "leakage" [15, 16]. In *Eunicella papillosa*, 60% to 70% of

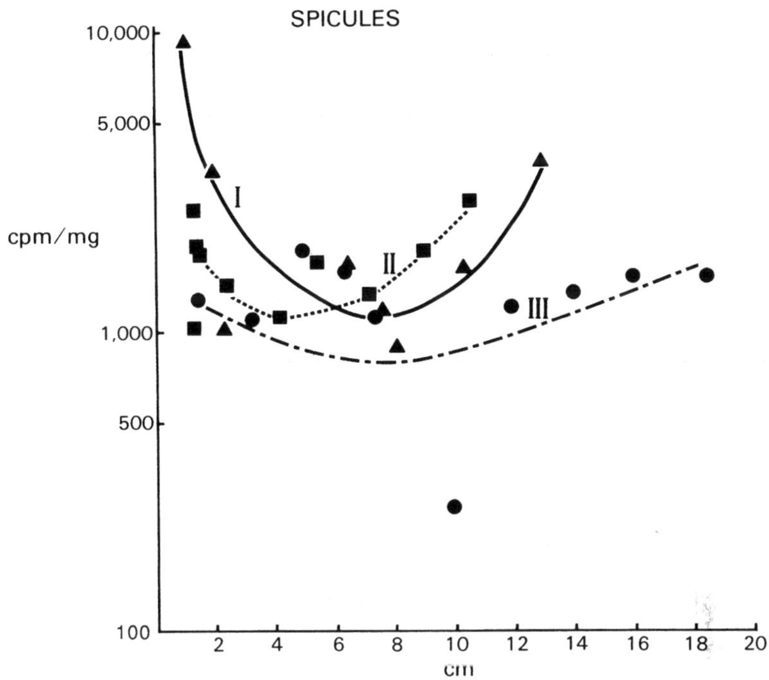

Fig. 9. ^{45}Ca uptake in spicules at different portions along length of branches of Colonies *I*, *II*, and *III*. *Ordinate*: Count · min^{-1} · mg^{-1} tissue. *Abscissa*: Distance from branch tip toward base. *Leptogorgia virgulata*. (With permission from [17])

the ^{45}Ca taken up by the animal remains in the tissue and is not deposited in the spicules [15]; 45% of the tracer which is taken up is returned to the water.

Rate of Calcification

In *Leptogorgia virgulata* spicule formation occurs throughout the animal colony and is directly related to the cellular activities of the branches [17]. In the branch tips, the degree of organic matrix formation and cell division are the highest, as indicated by the highest uptake of ^{3}H-aspartic acid and ^{3}H-thymidine. The rate of calcification is the highest in these regions and decreases to the mid-portion of colonies, followed by an increase toward the bases (Fig. 9). This latter rise is due to the increase in size and density of spicules which causes an increase in isotopic exchange. The rate of spicule formation is also the highest in the branch tips, but it decreases progressively in the lower branches in *Eunicella papillosa* [15].

Both spicule and axis weights progressively increase toward the basal portions in *Leptogorgia* [17] and *Eunicella* [3, 15]. Evidently, the branch increases in both length and diameter during growth with accompanying spicule forma-

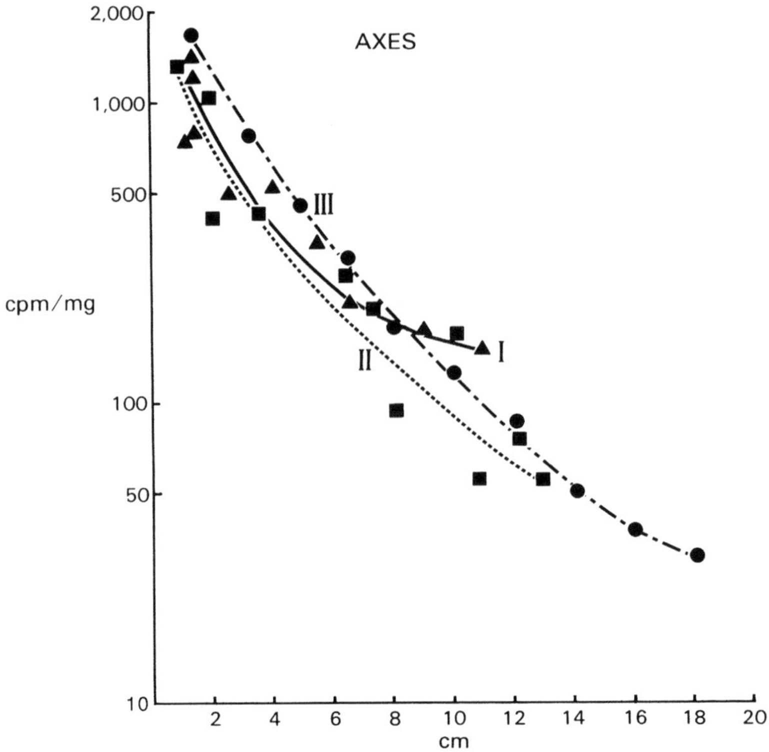

Fig. 10. ^{45}Ca uptake in axes at different portions along length of branches of Colonies *I*, *II*, and *III*. *Ordinate*: Counts · min^{-1} · mg^{-1} axis. *Abscissa*: Distance from branch tip toward base. *Leptogorgia virgulata*. (With permission from [17])

tion and axial growth. Thus, the spicule and axial weights become larger toward the base with increasing age of the branch, although the rate of calcification and cell division for both spicules and axes are lower at the base. Axes of apical regions show high rates of Ca uptake (Fig. 10) which indicates its active role in the calcification process.

Only few data are available on calcification rates of corals. In *Eunicella*, the average net calcification rate is 4.8 μg Ca mg $N^{-1} h^{-1}$ at 14°C [15]. The rates of ahermatypes range from 5.5 to 12.0 μg Ca mg $N^{-1} h^{-1}$ [18].

The Role of Carbonic Anhydrase in Spicule Formation

Carbonic anhydrase (CA) are associated with calcification in a number of invertebrates [19]. CA may affect calcification by catalyzing the conversion of CO_2 to HCO_3^- (equation 1), which would then be available for $CaCO_3$ formation (equation 2)

Table 1. ^{45}Ca uptake by control and Diamox-treated specimens. (From [19])

Sample	*Mean count min^{-1} mg^{-1} (n = 20)*	*SD*
Spicules		
10^{-6} M Diamox	764.57*	±78.37
10^{-4} M Diamox	760.03*	±75.87
Control	689.03	±89.32
Tissue		
10^{-6} M Diamox	596.03*	±326.88
10^{-4} M Diamox	830.70	±157.50
Control	841.47	±231.04
Axis		
10^{-6} M Diamox	2331.20*	±734.00
10^{-4} M Diamox	2126.50*	±850.90
Control	1615.00	±385.30

* Significantly different from control

$$CO_2 + H_2O \rightleftharpoons HCO_3^- + H^+. \quad (1)$$

$$Ca^{2+} + HCO_3^- \rightleftharpoons CaCO_3 + H^+. \quad (2)$$

CA may also catalyze the conversion of HCO_3^- to CO_2, the reverse reaction of Eq. (1), which would reduce the acidity of the calcifying milieu and provide a more conducive environment for $CaCO_3$ deposition. Another function assigned to CA is that of promoting decalcification. This enzyme may catalyze the reaction (1) to the right, producing protons and causing dissolution of calcareous structures.

The mean activity of CA in *Leptogorgia virgulata* is 2.7 nmol $\mu g^{-1} s^{-1}$, which is comparable to the values of crab gills and epidermis at intermolt and of some tissues of several molluscan species [18]. The activity in the polyps of the scleractinian coral *Pocillopora damicornis* is 0.162 nmol $\mu g^{-1} s^{-1}$ [20, 21]. Histochemically, intense CA activity is observed on the limiting membrane of the spicule-forming vacuole of *Leptogorgia* [19]. The membrane of the desmocyte microvilli, which is the closest cellular structure to the calcareous granules of the axis, also strongly exhibits this enzyme activity.

The function of CA in *Leptogorgia* spicule formation was investigated by a ^{45}Ca uptake experiment in the presence of the CA inhibitor Diamox [19]. In contrast to the results in other invertebrate calcification, treatment with Diamox causes an increase instead of a decrease in ^{45}Ca uptake in the spicules and axes (Table 1). This increase is interpreted to be related to the dynamics of HCO_3^- on the membranes adjacent to the calcifying structures. Assuming that HCO_3^- is not a limiting factor, as would be expected in a marine organism, CA may be involved in regulating this abundant ion in the spicule-forming vacuole and axis. CA may catalyze reaction (1) to the left and provide a means to regulate the amount of HCO_3^- available to react with Ca^{2+}. Thus, inhibition of CA would result in the increased availability of HCO_3^- for reaction with Ca^{2+}, thereby enhancing calcium carbonate precipitation.

It is possible that the CA in *Leptogorgia* is involved in ion regulation as in other aquatic invertebrates. Some CA, such as that on the plasma membrane of the scleroblasts, may catalyze the hydration of metabolic CO_2, generating H^+ and HCO_3^-. These ions could act as counter ions of Na^+ and Cl^-. The CA in this organism may also be involved in such functions as diffusion of CO_2 through membranes and maintenance of H^+ gradients across membranes [22]. The inhibition of CA involved in such processes may indirectly affect ^{45}Ca uptake. A priori, inhibition of such processes could potentially alter the physiological status of the organism, consequently altering the calcification process.

Hormonal Control of Calcification

Little is known concerning the hormonal control of calcification in octocorals. However, diiodotyrosine is found in gorgonians and iodine occurs in the axis [23]. Administration of thyroxine reduces the number of statoliths in the rhopalia of developing jellyfish [24]. Tissues of *Leptogorgia virgulata* stains positively for neurosecretory activity with paraldehyde fuchsin and lead hematoxylin. Immunofluorescence and immunogold labeling studies indicate that thyroxine or a thyroxine-like substance (TX) is present in the polyp vacuoles, gastrodermal tubes and nuclei, and electron dense bodies and spicules of scleroblasts of this animal. It is possible that TX is formed in the polyp and is transferred, by way of the gastrodermal tubes, to the scleroblasts, where it is involved in spicule formation (M. Corcoran and R.J. Kingsley, unpublished work).

Organic Matrix of Calcified Skeletons

Organic matrices of gorgonian spicules are approximately 1%–5% of the total weight of the spicules and contain lipids, carbohydrates, and proteins [25–27]. Quantitative data on those components are few. In *Pseudoplexaura flagellosa*, more than one-half of the weight of the matrix consists of lipids, about one-third is proteins, and the remainder is carbohydrates [7] (Table 2). In *Eugorgia ampla*, lipids are approximately 0.6% of the weight of spicules, or 60% of the organic matrix [26]. The data indicate that the lipids, followed by proteins are the major constituents of the gorgonian matrix.

The role of lipids in spicule formation is not well understood. However, long-chain fatty acids have been found to be responsible for the color of spicules and, accordingly, for the orange-yellow color of colonies of *Eugorgia ampla* [26]. One of these fatty acids is the "eugorgiaenoic acid" named by Fox et al. [26]. This is a yellowish, chemically unstable, non-aromatic, polyene-chained acid, which exhibits chemical and spectral properties suggestive of a

Table 2. Proximate analysis of *Pseudoplexaura* spicule matrix. (From [7])

Undecalcified spicules (n = 15)	*% Undecalcified spicule wt*	*% Total organic*	*% Insoluble organic*	*% Soluble organic*
Organic carbon	4.5 ± 0.4	100	88 ± 3.0	12.7 ± 2.0
Protein[a]	1.5 ± 0.2	33.3 ± 4.4	27 ± 4.6	4.3 ± 2.0
Carbohydrate[b]	0.4 ± 0.2	8.9 ± 0.4	0.5 ± 0.10	8.4 ± 1.0
Lipoidal pigment	2.6 ± 0.4	57.8 ± 8.9	55 ± 8.9	0

[a] Estimated by both amino acid analysis and fluorescence assay
[b] Estimated by phenol-sulfuric acid assay, using a glucose standard and by gas chromatography

chemical relationship with dihydrobixin. This pigment is conjugated to divalent cations, such as Ca and Mg in situ within the spicules.

Carbohydrates in the spicules of *Pseudoplexaura* include glucose and glucosamine which occupy nearly 60% of the total carbohydrate, followed by galactose and galactosamine (35%) and mannose (5%) [7]. Glucosamine or chitin is also present in *Leptogorgia virgulata* [28, 29]. Chitinous spicules are found in the sea pen, *Leioptilus guerneyi* [30].

The organic matrix of many invertebrates is known to consist of EDTA-soluble and -insoluble fractions. The insoluble matrix of the *Leptogorgia* spicules is usually fibrous, and the fibers are approximately 30 nm in diameter, displaying no periodic banding (Fig. 3). The autoradiographic investigation [31] using ^{3}H-aspartic acid as the tracer revealed that the protein moiety of the organic matrix is synthesized in the RER during the 5-min period following the 10-min incubation of the animal in the tracer. Within 5 min thereafter, the label moves from the RER to the Golgi complexes where the carbohydrate moiety is presumed to be synthesized. The label can be seen from 60 to 120 min in the spicules in the Golgi vacuole [31]. The entire process of the matrix synthesis and transport into the spicule takes approximately 2 h at 21°C in the summer.

Unlike vertebrate osseous and dental tissues made of calcium phosphates, invertebrate hard tissues of calcium carbonate rarely contain collagenous components in the insoluble matrices. The spicules of *Leptogorgia virgulata* are unique in that the major constituent of the insoluble matrix protein is a collagen without any periodic banding, and that the collagen content changes with the season [9]. As shown in Table 3 (columns A and C) the insoluble matrix of spicules collected in the summer months exhibits amino acid compositions typical of a collagen, i.e., having significant amounts of Hyp (Hydroxyproline), Hyl (Hydroxylysine) and 30%–33% Gly (Glycine). The composition is closer to that of the mesogleal collagen of sea anemone (Table 3, column I) than to the bovine bone collagen (Table 3, column J). However, the *Leptogorgia* spicule collagen contains less Hyp and more acidic amino acids than the sea anemone collagen. In fact, the *Leptogorgia* collagen is different in its amino acid composition from any other collagens reported in the cnidaria [29].

Table 3. Amino acids compositions of matrix protein fractions from isolated calcite spicules of *Leptogorgia virgulata*. (From [9])

	Residues per 1000 total residues									
	A TIM S.C.	B TIM 3/87	C TIM 7/87	D TIM 12/87	E Pepsin sol.	F Pepsin res.	G SM 12/87	H SM 7/87	I Sea anemone collagen[d]	J Bovine bone[e]
Hyp	74	8	82	31	80	11	1		103	98
Asp	93	132	76	130	84	153	516	577	73	45
Thr[a]	36	57	34	48	32	63	18	16	37	17
Ser[a]	43	57	42	53	43	78	16	13	41	34
Glu	109	87	94	92	112	76	34	32	97	74
Pro	66	67	65	59	64	50	17	19	67	123
Gly	295	158	332	225	335	161	210	185	339	337
Ala	83	78	82	76	77	67	130	108	65	109
Val	27	58	26	37	22	55	17	16	25	20
Cys[b]/2	2	16	1	10		17				
Met[c]	4	11	7	10	5	10				5
Ile	18	38	15	26	13	46	6	5	22	11
Leu	27	53	26	42	23	58	7	6	30	25
Tyr	8	24	8	20	6	32		2	3	4
Phe	9	31	9	22	7	35	4	4	8	13
His	3	17	4	11	3	11	2	1	1	4
Hyl	33	7	34	15	34	6			25	6
Lys	15	44	12	33	10	33	12	9	16	26
Arg	56	55	50	53	50	39	8	5	68	50

TIM, Total insoluble matrix; S.C., summer collection in South Carolina; *3/87, 7/87, 12/87*, dates of collection in North Carolina; *Pepsin sol.*, 7/87 collection, pepsin soluble material; *Pepsin res.*, 7/87 collection, pepsin insoluble material; *SM*, soluble matrix

[a] Uncorrected for hydrolysis

[b] Half Cys, sum of cysteic acid and cystine

[c] Sum of methionine sulphoxide and methionine

[d] [37]

[e] [38]

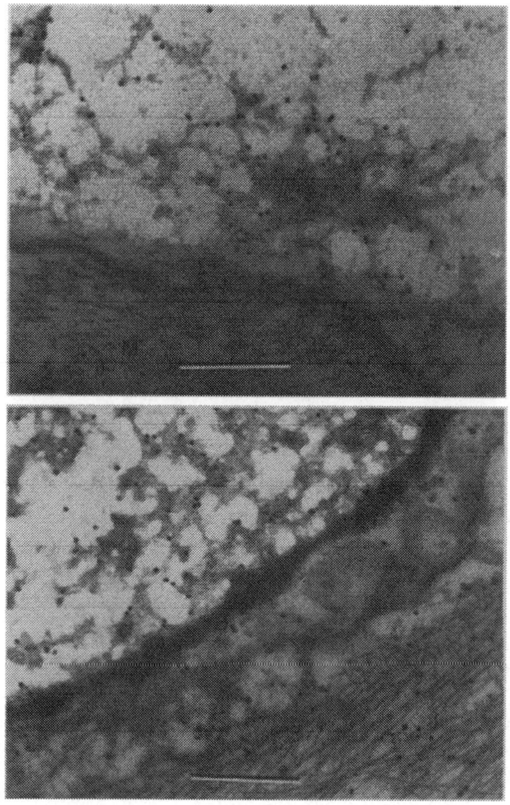
a
b

Fig. 11. **a** Immunogold labeling for the collagenous 140 kd protein in the summer spicule of *Leptogorgia virgulata*. The labels are on the framework of the insoluble matrix. Mesoglea fibers (*M*) are scarcely labeled. *Scale bar* = 1 μm. **b** Immunogold labeling for the 40 kd non-collagenous protein in the summer spicule of *Leptogorgia virgulata*. The labels are at the periphery of the insoluble matrix and on the mesoglea fibers. *Scale bar* = 1 μm

The collagenous nature of the spicule insoluble matrix has also been confirmed by the presence of the intermolecular reducible cross-link, dihydroxylysinonorleucine [9]. The collagen content decreases in the December samples, as indicated by the lower values of Hyp (31/1000), Hyl (15/1000), and Gly (225/1000) (Table 3, column D). The values further decrease in the March sample, showing very little of the collagenous characteristics (Table 3, column B). This composition is similar to that of the non-collagenous minor component of the summer insoluble matrix (Table 3, column F). These results are substantiated by SDS-PAGE and immunocytochemical findings [29] (Watabe N et al. (1992) unpublished material). SDS-PAGE patterns of the summer insoluble matrix consist of protein bands at 500 kd, 140 kd, 120 kd, and 40 kd, of which the 140 kd and probably 120 kd proteins are collagenous. The spring samples have only the 40 kd protein band, but are sometimes accompanied by very faint bands of 140 kd and120 kd.

In the summer spicules, the collagenous 140 kd protein is detected within the insoluble matrix framework by immunogold labeling with antibodies against this protein (Fig. 11.a). The mesogleal fibers surrounding the scleroblasts and spicules show very little labeling. On the other hand, the labels for non-

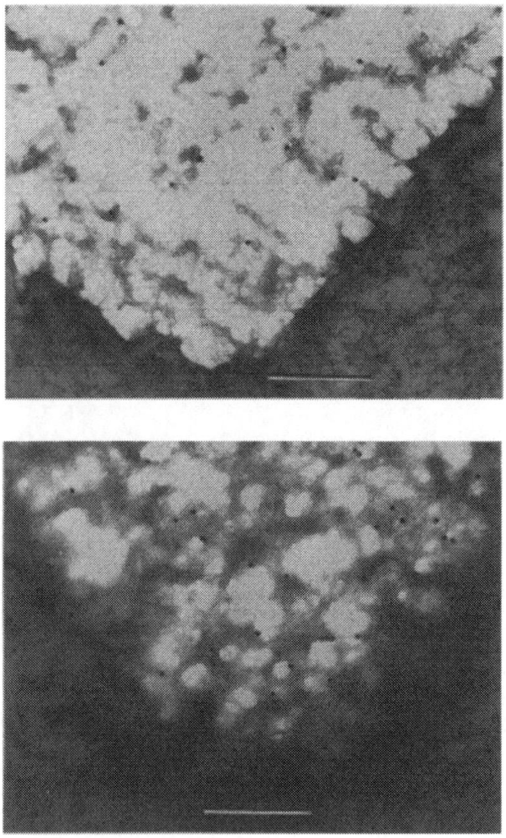
a

b

Fig. 12. a Matrix of the winter spicule of *Leptogorgia virgulata* showing few immunogold labels for the 140 kd collagenous protein. *Scale bar* = 1 μm. **b** Winter spicule matrix and mesoglea of *Leptogorgia virgulata* labeled with immunogold for the non-collagenous 40 kd protein. *Scale bar* = 1 μm

collagenous 40 kd insoluble matrix protein are at the periphery of the insoluble matrix framework, and the mesogleal fibers show a heavy concentration of immunogold which labels antibodies against this protein (Fig. 11.b). In the winter/spring spicules, both the insoluble matrix framework and the mesogleal fibers are scarcely labeled for the 140 kd collagen (Fig. 12.a), whereas the periphery of the insoluble matrix framework and the mesogleal fibers are labeled heavily for the non-collagenous 40 kd protein (Fig. 12.b). However, the 40 kd labels are not as heavy as those in the summer.

In contrast to the insoluble matrix, the biochemical characteristics of the soluble matrix do not change throughout the year. The matrix is very high in aspartate, approximately 500 to 650 residues per thousand residues, glycine, and alanine [25] (Table 3, columns G and H). A highly acidic calcium-binding glycoprotein is present in this matrix [28], and SDS-PAGE shows that the apparent molecular weight of this protein is 88,000 [29]. In addition, approximately 10 minor protein bands ranging from 12 kd to 100 kd were detected in 12% gels of both the summer and winter samples [29]. Similar to the labels for the 40 kd insoluble protein, the immunogold marker for the soluble 88 kd glycoprotein decorates the periphery of the insoluble matrix framework (Fig.

Fig. 13. Immunogold particles labeling the 88 kd soluble protein at the periphery of the insoluble matrix. *Leptogorgia virgulata*. *Scale bar* = 0.5 μm

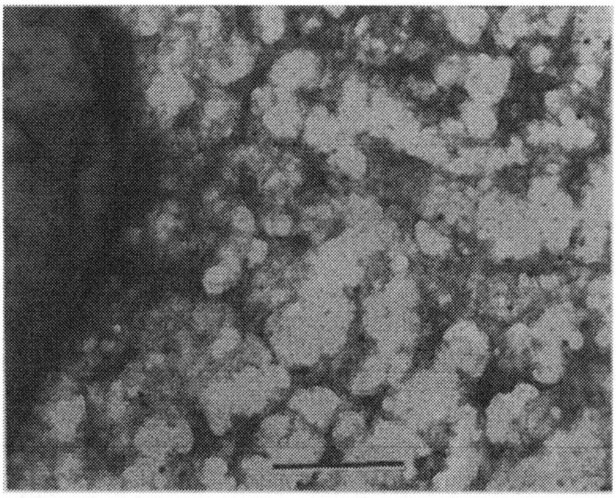

13). The spatial relationship between the 88 kd soluble and 40 kd insoluble proteins is unknown.

The results of the biochemical and immunocytochemical investigations indicate that the content of the spicule collagen is very much reduced in the winter/spring animals, but that of the non-collagenous components is not. This implies that the collagen component within the spicules becomes degraded and digested during the fall/winter months, which is preceded by (partial) spicule decalcification. At present, no information is available concerning the process of degradation and digestion of this collagen. It may involve activation of some enzymes, such as a specific collagenase, by certain seasonal cues. Spicule decalcification may result from shifts in the activities of some enzymes, e.g., lysosomal enzymes, carbonic anhydrase (CA), etc. In fact, these enzymes have been cytochemically localized in the scleroblasts [4, 19]. By some seasonal cue, CA may be activated to catalyze the reaction $CO_2 + H_2O \rightarrow HCO_3^- + H^+$. The H^+ react with $CaCO_3$ and decalcify spicules ($CaCO_3 + H^+ = Ca^{2+} + HCO_3^-$). The activation may be hormone dependent. Lysosomal enzymes may react with spicules and dissolve crystals. The enzymes may also digest degraded collagen. It is known that lysosomal enzymes are affected by thyroid hormone [32], and substances which immunologically cross-react with thyroxine (T4) have been localized in *Leptogorgia* (see Hormonal Control of Calcification). Furthermore, an inhibition of T4 activities has been shown to affect the calcification in this organism (R.J. Kingsley and N. Watabe, unpublished work). Thus, it is conceivable that season-dependent activities of hormone or hormone-like substances and enzymes regulate collagenous (and non-collagenous) organic matrix formation and degradation as well as spicule calcification and decalcification in *Leptogorgia*.

Functions of the organic matrix in spicule formation are not clear. Acidic glycoprotein and/or proteoglycan macromolecules are generally considered to influence calcium carbonate crystal nucleation and growth in vitro. As for the spicule matrix, information is lacking concerning the molecular organization of various protein fractions. However, immunocytochemical evidence shows that the collagen is the backbone of the organic matrix framework of the summer spicule in *Leptogorgia virgulata*, and that the non-collagenous insoluble and soluble fractions are attached to it. The fact that the collagenous component occurs only in the summer when the spicule calcification is very active suggests that this spicule collagen is essential for the initial calcite nucleation. In fact, the insoluble matrix obtained after decalcification of spicules of *Leptogorgia* has been shown to recalcify in vitro [33]. Nevertheless, this does not necessarily mean that the collagenous insoluble matrix itself is the calcite nucleator in vivo. Although this matrix is calcifiable, it is covered by the soluble matrix and may not be able to attract Ca^{2+} ions. Recently, Crenshaw [34] and Gunthorpe et al. [35] presented a theory and evidence for the role of organic matrix in calcification in vitro. They have shown that polyanions, such as those of the soluble matrix, induce nucleation when immobilized on an insoluble substrate. No nucleation occurs when those polyanions are in solution. This scheme may be applicable to the *Leptogorgia* crystal nucleation. The soluble matrix attached to the collagenous insoluble matrix may be the nucleator and the collagenous matrix may regulate crystal growth. In *Veretillum cynomorium*, the axial rod contains calcite crystals through which collagen fibers penetrate, but the collagen fibers are never impregnated with the calcite. Ledger and Franc consider that the collagen is not responsible for the initial nucleation in this organism [11]. More data are necessary to determine the roles of the insoluble and soluble matrices in spicule formation in vivo.

Nothing is known concerning the morphogenesis of the spicules. The organic matrix (insoluble) deposited initially in the spicule vacuole of *Leptogorgia virgulata* does not assume the final spicule morphology. This implies that the final, species-specific and symmetrical shape of spicules (not of the individual crystals) is not likely to be determined by the initial insoluble matrix morphology. The form of the initial spicule is a rather smooth-surfaced rod without warts and decorations [4]. Thus, the basic form of the spicule may be the rod, which is presumably determined solely by the insoluble matrix substrate to which the soluble matrix is attached, or by combined effects of crystal growth and the substrate. The spicule continues to grow into a larger rod or spindle by further alternate deposition of the matrix and crystal layers. After attaining a certain size, symmetrical modification of the spicule form occurs, resulting in the elaborate and yet systematic decorations of the spicule. Watabe and Kingsley [36] proposed that the presence of regularly spaced inhibitors of crystal growth in the vacuole may be one of the factors controlling this morphogenesis. Patterns of distribution of those inhibitors along the vacuole membrane may be species specific. Biochemical, physiological, cytochemical, and ultrastructural analyses of those spicule-forming vacuoles and vacuole membranes may help to elucidate the morphogenesis of spicules as well as certain other biominerals.

Acknowledgements. All the authors' gorgonian research mentioned herein have been supported by the National Science Foundation grants #PCM-

8201389, DCB-8502689, DCB-880109, and partially by DIR-8805143, and by the funds of the Electron Microscopy Center, The University of South Carolina. We thank Betty Bynum and Robert Tye Whitaker for assistance in the preparation of the manuscript.

References

1. Bayer FM, Harding BO (1968) The free-living lower invertebrates. Macmillan, London, pp 229
2. Vinogradov AP (1953) The elementary chemical composition of marine organisms. Mem Sears Fdn mar Res No. 2, New Haven, pp 647
3. Velimerov B, Bohm EL (1976) Calcium and magnesium carbonate concentrations in different growth regions of gorgonians. Mar Biol 35:269–275
4. Kingsley RJ, Watabe N (1982) Ultrastructural investigation of spicule formation in the gorgonian *Leptogorgia virgulata* (Lamarck) (Coelenterata: Gorgonacea). Cell Tissue Res 223:325–334
5. Dunkelberger DG, Watabe N (1974) An ultrastructural study on spicule formation in the Pennatulid colony *Renilla reniformis*. Tissue Cell 6:573–586
6. Goldberg WM, Benyahu Y (1987) Spicule formation in the gorgonian coral *Pseudoplexaura flagellosa* 1. Demonstration of intracellular and extracellular growth and the effect of Ruthenium red during decalcification. Bull Mar Sci 40:287–303
7. Goldberg WM (1988) Chemistry, histochemistry and microscopy of the organic matrix of spicules from a gorgonian coral. Relationship to Alcian blue staining and calcium binding. Histochemistry 89:163–170
8. Kingsley RJ, Bernhardt AM, Wilbur KM, Watabe N (1987) Scleroblast cultures from the gorgonian *Leptogorgia virgulata* (Lamarck) (Coelenterata: Gorgonacea). In Vitro Cell Dev Biol 23:297–302
9. Kingsley RJ, Tsuzaki M, Watabe N, Mechanic GL (1990) Collagen in spicule organic matrix of the gorgonian *Leptogorgia virgulata*. Biol Bull 179:207–213
10. Chia FS, Crawford B (1977) Comparative fine structural studies of planulae and primary polyp of identical age of the sea pen, *Ptilosarcus gurneyi*. J Morphol 151:131–158
11. Ledger PW, Franc S (1978) Calcification of the collagenous axial skeleton of *Veretillum cynomorium* Pall. (Cnidaria: Pennatulacea). Cell Tissue Res 192: 249–266
12. Kingsley RJ, Watabe N (1984) Calcium uptake in the gorgonian *Leptogorgia virgulata*. The effects of ATPase inhibitors. Comp Biochem Physiol [A] 79:487–491
13. Kingsley RJ, Watabe N (1985) An autoradiographic study of calcium transport in spicule formation in the gorgonian *Leptogorgia virgulata* (Lamarck) (Coelenterata: Gorgonacea). Cell Tissue Res 239:305–310
14. Kingsley RJ, Watabe N (1985) Ca-ATPase localization and inhibition in the gorgonian *Leptogorgia virgulata* (Lamarck) (Coelenterata: Gorgonacea). J Exp Mar Biol Ecol 93:157–167
15. Velimirov B, King J (1979) Calcium uptake and net calcification rate in the octocoral *Eunicella papillosa*. Mar Biol 50:349–358
16. Barnes DJ, Crossland CJ (1972) Coral calcification: Sources of error in radioisotope techniques. Mar Biol 42:119–129

17. Kingsley RJ, Watabe N (1989) The dynamics of spicule calcification in whole colonies of the gorgonian *Leptogorgia virgulata* (Lamarck) (Coelenterata: Gorgonacea). J Exp Mar Biol Ecol 133:57–65
18. Goreau TF (1961) On the relation of calcification to primary productivity in reef-building organisms. In: Lenhoff HM, Loomis WF (eds) The biology of *Hydra*. University of Miami Press, Miami, pp 269–285
19. Kingsley RJ, Watabe N (1987) Role of carbonic anhydrase in calcification in the gorgonian *Leptogorgia virgulata*. J Exp Zool 241:171–180
20. Graham D, Smillie RM (1976) Carbonate dehydratase in marine organisms of the Great Barrier Reef. Aust J Plant Physiol 3:113–119
21. Sikes CS, Wheeler AP (1982) Carbonic anhydrase and carbon fixation in coccolithophorids. J Phycol 18:423–426
22. Jones WC, Ledger PW (1986) The effect of diamox and various concentrations of calcium on spicule secretion in the calcareous sponge *Sycon ciliatum*. Comp Biochem Physiol [A] 84:149–158
23. Gorbman AM, Clements M, O'Brien R (1959) Utilization of radioiodine by invertebrates, with special study of several annelida and mollusca. J Exp Zool 127:75–89
24. Spangenberg DB (1984) Effects of exogenous thyroxine on statolith synthesis and resorption in *Aurelia*. Am Zool 24:917–924
25. Kingsley RJ, Watabe N (1983) Analysis of proteinaceous components of the organic matrices of spicules from the gorgonian *Leptogorgia virgulata*. Comp Biochem Physiol [B] 76:443–447
26. Fox DL, Smith VE, Grigg RW, MacLeod WD (1969) Some structural and chemical studies of the microspicules in the fan-coral *Eugorgia ampla* Verrill. Comp Biochem Physiol 28:1103–1114
27. Silberberg MS, Ciereszko LS, Jacobson RA, Smith EC (1972) Evidence for a collagen-like protein within spicules of coelenterates. Comp Biochem Physiol [B] 43:323–332
28. Samata T, Kingsley RJ, Watabe N (1989) Ca-binding glycoprotein from the spicules of the octocoral *Leptogorgia virgulata*. Comp Biochem Physiol [B] 94:651–654
29. Watabe N, Oishi M, Kingsley RJ (1991) The organic matrix of spicules of the gorgonian *Leptogorgia virgulata*. In: Suga S, Nakahara H (eds) Mechanisms and phylogeny of mineralization in biological systems. Springer, Tokyo, pp 9–16
30. Shapeero W (1969) A positive chitosan test for spicules in the Anthozoan order, Pennatulacea. Pacif Sci 23:261–263
31. Kingsley RJ, Watabe N (1984) Synthesis and transport of the organic matrix of the spicules of the gorgonian *Leptogorgia virgulata* (Lamarck) (Coelenterata: Gorgonacea) An autoradiographic investigation. Cell Tissue Res 235:533–538
32. Gaton DD, Gaton E, Wolman M (1987) Are the effects of thyroid hormone on target organs mediated through lysosomal enzymes? Cell Mol Biol 33:619–624
33. Watabe N, Berhnardt AM, Kingsley RJ, Wilbur KM (1986) Recalcification of decalcified spicule matrices of the gorgonian *Leptogorgia virgulata* (Cnidaria: Anthozoa). Trans Am Microsc Soc 105:311–318
34. Crenshaw MA (1990) Mineral induction by immobilized polyanions. In: Suga S, Nakahara H (eds) Mechanisms and phylogeny of mineralization in biological systems. Springer, Tokyo, pp 101–105
35. Gunthorpe ME, Sikes CS, Wheeler AP (1990) Promotion and inhibition of calcium carbonate crystallization in vitro by matrix protein from blue crab exoskeleton. Biol Bull 179:191–200

36. Watabe N, Kingsley RJ (1990) Extra-, inter-, and intracellular mineralization in invertebrates and algae. In: Crick RE (ed) Origin, evolution, and modern aspects of biomineralization in plants and animals. Plenum, New York, pp 209–223
37. Norwack H, Nordwig A (1974) Sea-anemone collagen: Isolation and characterization of the cyanogen-bromide peptides. Eur J Biochem 45:333–342
38. Herring GM (1972) The organic matrix of bone. In: Bourne GH (ed) The biochemistry and physiology of bone structure, vol 1. Academic, New York, pp 127–189

Involvement of Mammalian Insulin and Insulin-like Peptides in Shell Growth and Shell Regeneration in Molluscs

Abu S.M. Saleuddin, Venkatalakshmi M. Sevala, Veeresh L. Sevala, Spencer T. Mukai, and Hamid R. Khan[1]

Summary. A brief review of the neuroendocrine control of shell formation/shell regeneration is presented. In *Helisoma duryi*, neurosecretory mediodorsal cells (MDC), which occur in 2 major groups in the two cerebral ganglia, are coupled via gap junctions, as revealed by electron microscopy and Lucifer yellow injections. The axons of MDC terminate in the two median lip nerves where neurosecretory granules are released by exocytosis. Extracellular electrical recordings from the median lip nerve, the neurohemal area of MDC, show that MDC are active in non-reproducing virgin snails and less active in reproducing snails. Many MDC are marked by porcine insulin antiserum, suggesting that they produce insulin-like peptide(s) (ILP). ILP and porcine insulin stimulate periostracal protein synthesis in vitro in the mantle. Immunoblotting experiments suggest the presence of two ILP immunoreactive bands in the hemolymph of *Helisoma* and one band in *Otala*. When a piece of shell is removed from the edge in *Helisoma*, shell regeneration is faster in fast growing snails than in slow growing ones. However, ILP titers are highest in slow growing snails and are generally lower at the peak of shell regeneration. ILP titers in the hemolymph fluctuate during a 24 h period, and are influenced by photoperiods. In *Helix aspersa*, microgram quantities of bovine insulin stimulate growth.

Key words: Insulin-like peptide—Shell growth—Molluscs—Radioimmunoassay—Shell regeneration

Introduction

The molluscan shell, which offers protection from predators and from extreme environmental fluctuations, such as temperature and humidity, is constructed on an uniform plan. Typically, it consists of an outermost organic layer, the periostracum, which overlies a series of calcareous layers made primarily of $CaCO_3$ crystals. These inorganic layers are supported by an organic matrix. While the shell structure of many molluscs is built on an uniform plan, it's

[1] Department of Biology, York University, North York (Toronto), Ontario, Canada M3J 1P3

mode of formation is not. In many molluscs, shell growth is incremental while in some molluscs a daily rhythm of shell deposition can be detected. This incremental growth of the shell whether it be sub-daily, daily, weekly, monthly, annually, or irregular can be influenced by various internal as well as external factors [1]. Among the latter category, tides, salinity, and photoperiod have been implicated. It is known that photoperiod not only influences the shell increment rate but also the thickness of a specific shell layer [2]. Environmental conditions, such as temperature, also influence the types of crystal morphs of $CaCO_3$ to be deposited [1]. Various control mechanisms of shell formation in molluscs have been described wherein each of these must involve a complex series of cellular responses by the mantle in first secreting the periostracum and then the calcareous shell. In order to deposit the inorganic shell material, the mantle is involved in ionic transport of Ca^{2+} and HCO_3^- to the extrapallial space whereby mechanism(s), still not quite well understood, form ordered layers of $CaCO_3$ crystals on an organic matrix. Although a role of the central nervous system (CNS) in shell formation has been speculated upon for some time, experimental evidence was not forthcoming until Dillaman et al. [3] showed that a group of neurosecretory cells in the visceral ganglion of the CNS in *Helisoma duryi* exhibited cyclical fluctuation of neurosecretory material in response to shell damage. Earlier, Lubet [4] showed that a factor from the cerebral ganglia in the prosobranch *Crepidula fornicata* stimulated body growth.

When a piece of shell is removed from a mollusc, a new piece is regenerated in the area of injury. The rate of shell repair is variable from species to species and is also dependent upon the site of shell injury. For example, shell repair rate is high if the injury is at the shell edge of *Helisoma* but the rate is slow if the area of injury is distant to the edge [5, 6]. By contrast, the repair is faster in *Helix* if the injury is made distant to the shell edge [7, 8]. Structurally, the repaired shell can be identical to the normal shell if the injury occurs at the edge [5] or it can be quite different, not only structurally but also in mineralogy, when the injury is made distant to the edge, as in *Helix* [7, 9]. Both regular incremental shell growth and shell regeneration involve regular and rapid turnover of $CaCO_3$ and proteins, and these systems are thus amenable to experimental analysis when involvement of the CNS is sought. To address the question whether or not the CNS exerts a direct influence on shell formation and/or shell regeneration, we refer to the findings of Dillaman et al. [3]. These authors first established the neuroendocrine centers in the CNS of *H. duryi*, and then studied the effect of shell damage on the synthetic activity of neurosecretory cells in the CNS. They reported that a group of neurosecretory cells in the visceral ganglion showed cyclical activities of paraldehyde fuchsin (PAF)-positive materials, suggesting that these cells are involved in the shell repair process. These cells were later called Type-1 cells [10] and Saleuddin et al. [11] suggested that they are involved in calcium mobilization, a prerequisite for the shell repair process. Several reports have shown that the injection of suboesophageal ganglion extract increases hemolymph calcium levels in a number of pulmonates [12, 13].

Another line of evidence for the involvement of the CNS in shell formation in *H. duryi* came from the study of Kunigelis and Saleuddin [2] who showed that brain extracts from fast growing snails, when injected into slow growing snails, caused faster shell growth in the treated group. Furthermore, injections of either supra- or suboesophageal fractions of the CNS had a more pronounced effect on shell growth than whole CNS [11].

The involvement of neurosecretory cells in body and shell growth in *Lymnaea stagnalis* was attributed to one kind of cell located in 4 groups in the cerebral ganglia [14]. These cells, termed light green cells (LGC) which are homologous to mediodorsal cells (MDC) in *H. duryi*, are under an inhibitory control of two lateral lobes located within the cerebral ganglia [15]. However the exact cellular origin of this inhibition is not known. This inhibitory role was extrapolated from the observation that the surgical removal of both lateral lobes caused giant growth in snails and that their reimplantation restored normal growth [15]. The lateral lobes of *H. duryi* also exert an inhibitory effect on its MDC [16]. An alternate explanation for this inhibition of growth would be that lateral lobes stimulate the endocrine dorsal bodies involved in female reproduction, and their removal would thus inhibit reproduction. Because there is an antagonistic relationship between reproduction and growth, growth (shell) will accelerate when the reproduction is inhibited [17]. In a series of papers, Dogterom et al. showed that the growth factor from LGC in *L. stagnali* increased uptake of tritiated tyrosine and ^{45}Ca from the medium by the mantle edge. They detected a calcium-binding protein in the mantle edge whose concentration was effected by the removal of LGC [18–20]. Neurosecretory cells in the CNS of several other pulmonates are known to be involved in growth [21].

The influence of the CNS on in vitro periostracum formation and calcium transport was studied by Kunigelis and Saleuddin [16, 22] who showed that the addition of the extract of CNS from a fast growing snail increased periostracum formation in the mantle from a slow growing snail. Furthermore, the putative brain factor increased incorporation of certain amino acids, such as glycine. They also showed that the periostracum formed in vitro incorporated ^{45}Ca from the medium, and that the CNS from a fast growing snail increased the incorporation by the mantle from a slow growing snail. Finally, they found that the dorsal epithelium of the mantle edge was important for ^{45}Ca incorporation, and that the lateral lobes exerted an inhibitory influence both on periostracum formation and ^{45}Ca incorporation.

The LGC once thought to produce somatostatin [23] in *L. stagnalis* have now been shown to produce molluscan insulin-like peptides (MIPs), one of which has been sequenced [24]. In situ hybridization with cDNA probes for MIPs showed that the transcription of MIPs occurs in LGC and canopy cells [21]. MIP has been shown to have significant homology with insect prothoracicotropic hormone (PTTH) and mammalian insulin [25]. The effect of MIP and vertebrate insulin on shell growth in *Lymnaea* has not yet been demonstrated.

MDC Produce Insulin-like Peptide (ILP) in *H. Duryi*

The distribution and the fine structure of MDC have been described [17, 26]. There are about 30 MDC in each cerebral ganglion lying close to the dorsal bodies (DB) (Figs. 1–3). As in *L. stagnalis*, the axons of the MDC in *Helisoma* terminate in the median lip nerves which comprise the neurohemal area of the MDC (Figs. 1, 3). When the cells are injected with Lucifer yellow, it can be seen that a few axons of MDC also traverse the cerebral commissure. It is not known where these axons terminate (Figs. 1, 3). In *L. stagnalis*, Benjamin et al. [27] suggested that, while MDC in one cerebral ganglion are mostly coupled, they are not coupled with the contralateral cells. However when Ca^{2+} was replaced by Ba^{2+} in the bathing solution, synchronous bursting activities were recorded from both median lip nerves, which suggested that all MDC/LGC are coupled [28]. In *Helisoma*, axons of MDC from one cerebral ganglion were seen to pass toward contralateral cells. This raises the possibility of coupling between the ipsilateral and contralateral cells. Although *Helisoma* is a hermaphroditic pulmonate, it reproduces by cross fertilization. Reproducing animals grow more slowly than virgin and castrated snails [17]. Extracellular electrical recordings from the median lip nerve (the neurohemal area of MDC) showed very little activity in randomly mating snails, whereas recordings from virgin and castrated snails showed increased frequency of spontaneous action potentials (Fig. 4). This finding suggests an increased release of neurosecretory material from the axons of MDC located in the median lip nerves of virgins but not mated snails, further suggesting that the MDC are involved in the control of shell growth. The MDC in *Helisoma* were found to be about 45 μm in diameter and were full of neurosecretory granules measuring 150 nm in diameter (Fig. 5). Gap junctions were detected between the cell bodies as well as between axon endings of MDC (Figs. 3, 6, 7). Axons of the MDC terminated at the median lip nerve and were also filled with neurosecretory granules (Fig. 8). The axons showed release of neurosecretory granular content in the connective tissue by single or multiple exocytosis (Figs. 9–12). The rate of exocytosis by these cells was higher in virgins than in mated snails and was also higher during the night. It is worth nothing that

Fig. 1. Whole mount of *Helisoma* brain in which 10% Lucifer yellow was microinjected into an MDC (*large arrow*); the dye was subsequently transferred into two other MDC (*small arrows*); the dye-filled axons of MDC are seen in cerebral ganglion (*CG*), cerebral commissure (*CM*), and median lip nerve (*MLN*). *VG*, Visceral ganglion. *Bar* = 100 μm

Fig. 2. Cross-section of the cerebral ganglion (*CG*) of *Helisoma* showing indirect immunofluorescent anti-insulin staining of MDC. *DB*, Dorsal bodies. *Bar* = 30 μm

Fig. 3. Diagrammatic illustration of the supraesophageal portion of the CNS of *Helisoma* showing the *MDC* and the location of *gap junctions*. Axons of MDC do travel to contralateral sides but their final destinations are not known. Not drawn to scale

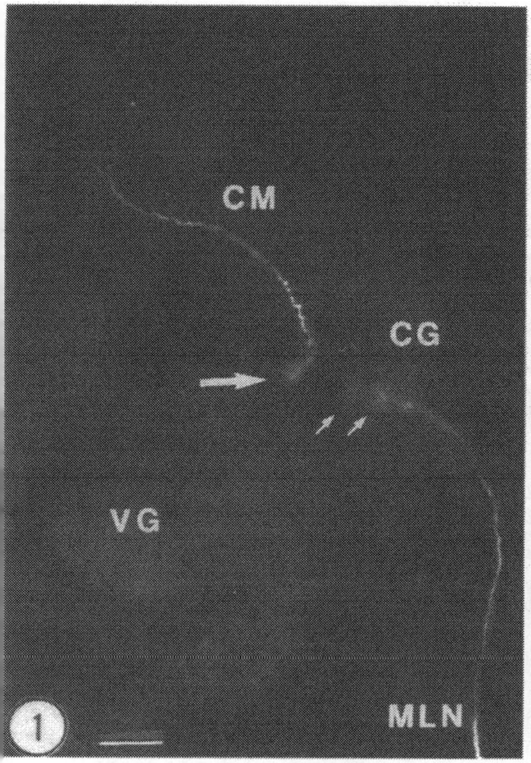
CM
CG
VG
MLN
1

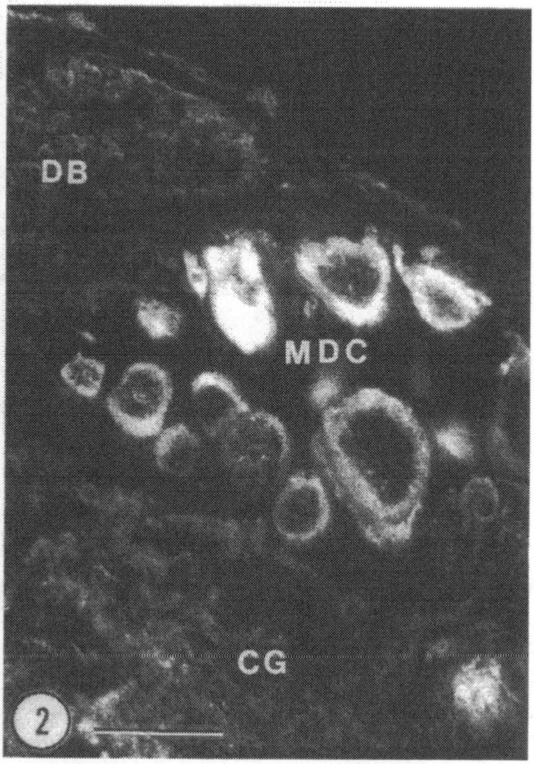
DB
MDC
CG
2

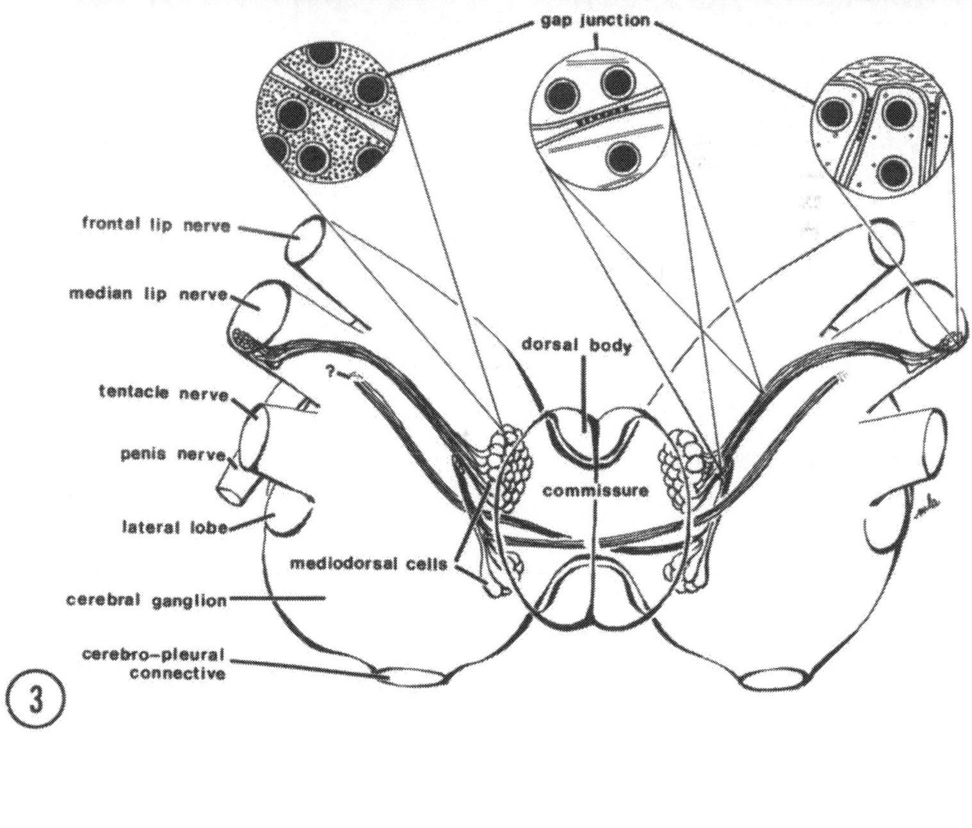
gap junction
frontal lip nerve
median lip nerve
dorsal body
tentacle nerve
penis nerve
commissure
lateral lobe
mediodorsal cells
cerebral ganglion
cerebro-pleural
connective
3

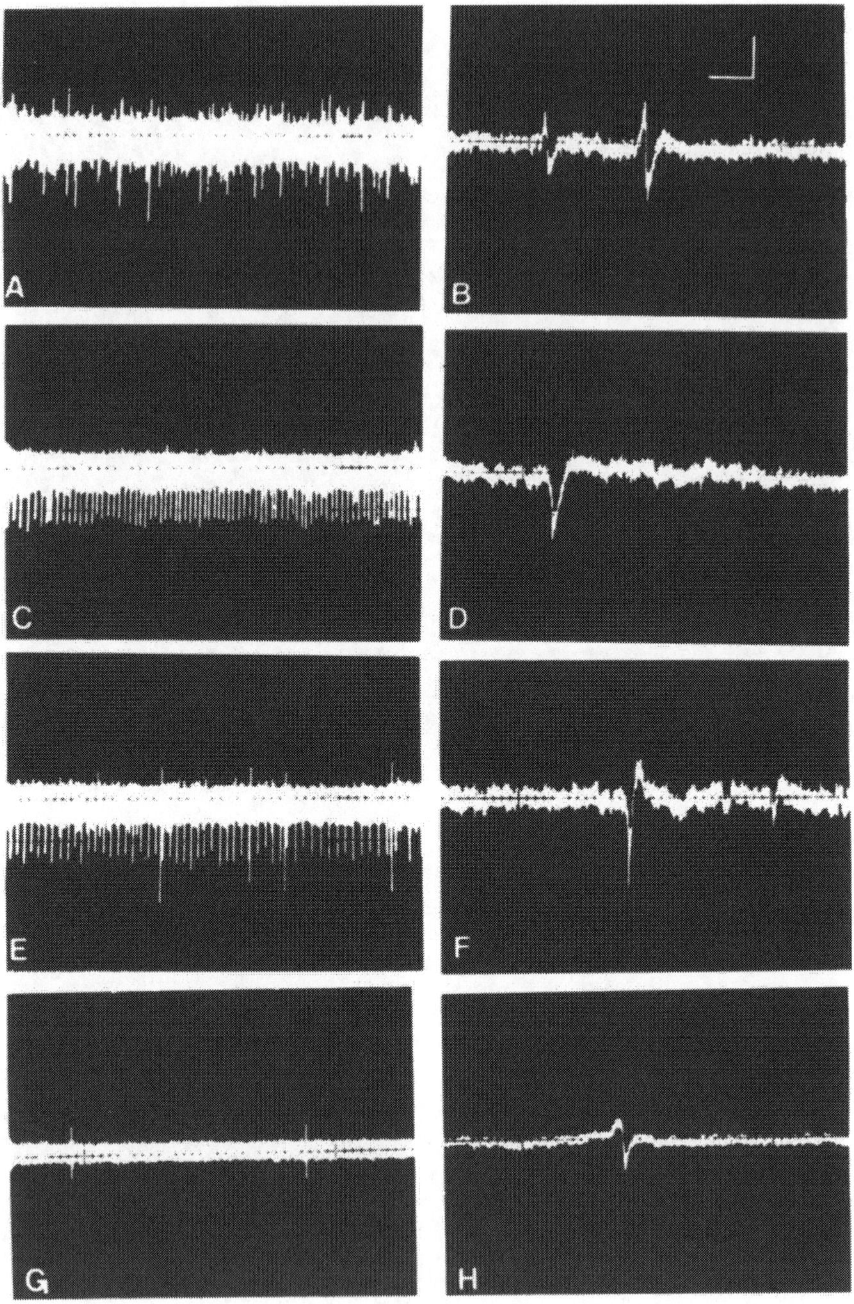

Fig. 4. Spontaneously occurring action potentials recorded extracellularly using suction electrodes from the median lip nerve of virgin (**A–D**) and castrated snails (**E–F**) (*Helisoma*). Little spontaneous activity was observed from the median lip nerve of randomly mating snails (**G–H**). The traces on the *right* are extensions of the time scale from the *left* traces. Scale mark: *left panel*, vertical: 100 μv, horizontal: 1s; *right panel*, vertical: 100 uv, horizontal 10 ms. (From [17] with permission)

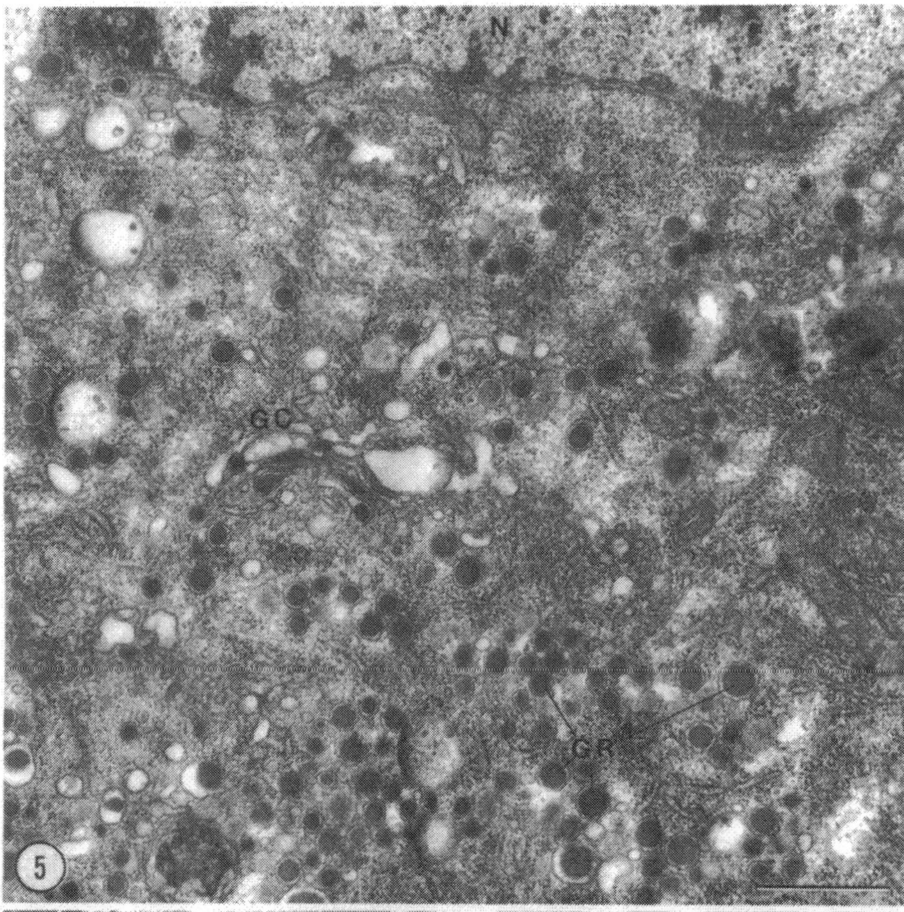

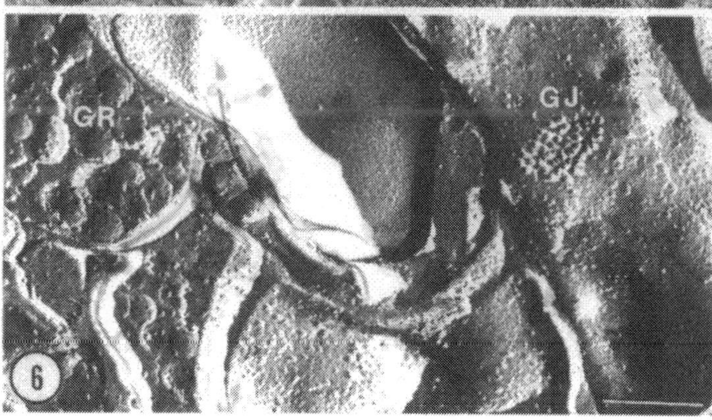

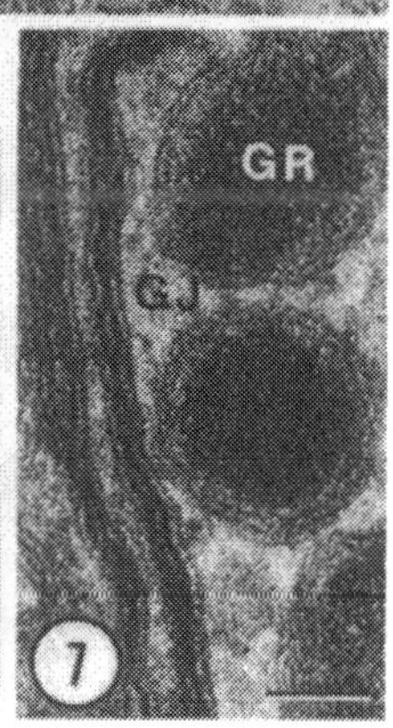

Fig. 5. Cross-section of MDC of a virgin snail showing Golgi complex (*GC*), elementary granules (*GR*), and nucleus. *Bar* = 0.5 µm

Fig. 6. Freeze-fracture replica of the basal portion of an MDC showing gap junctions (*GJ*) and elementary granules (*GR*). *Bar* = 0.2 µm

Fig. 7. *Helisoma*. Cross-section of MDC showing the gap junction (*GJ*) and elementary granules (*GR*). *Bar* = 0.02 µm

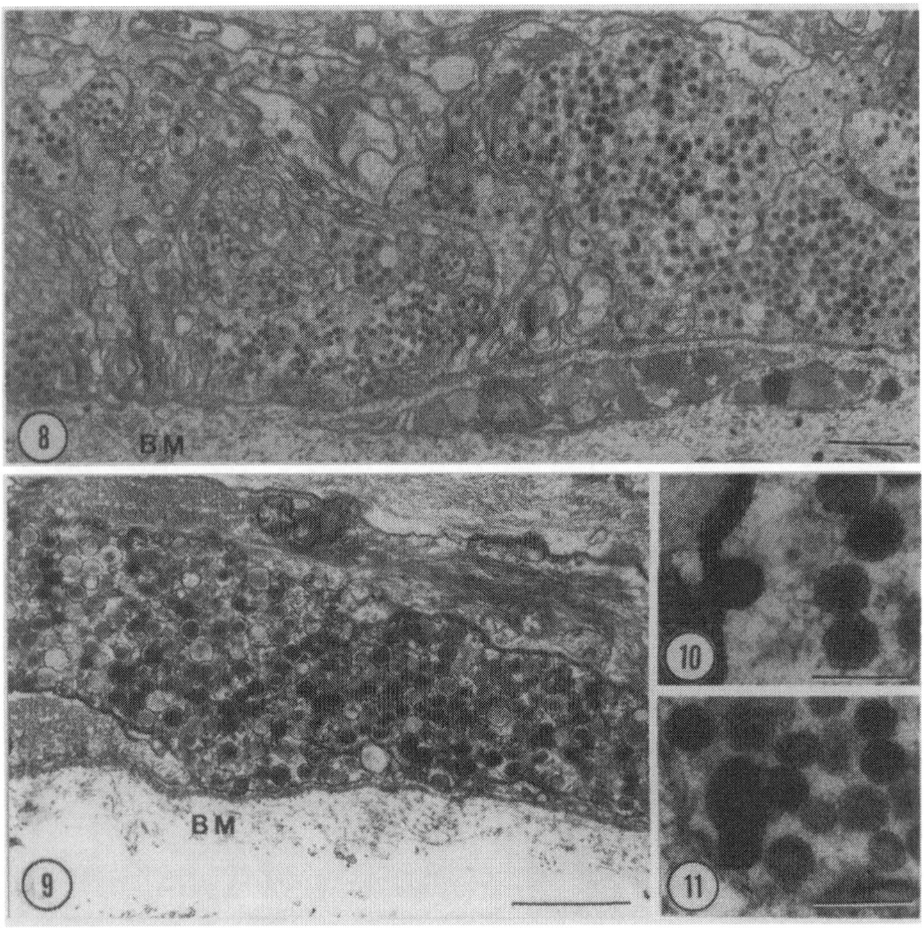
8
BM
9
BM
10
11

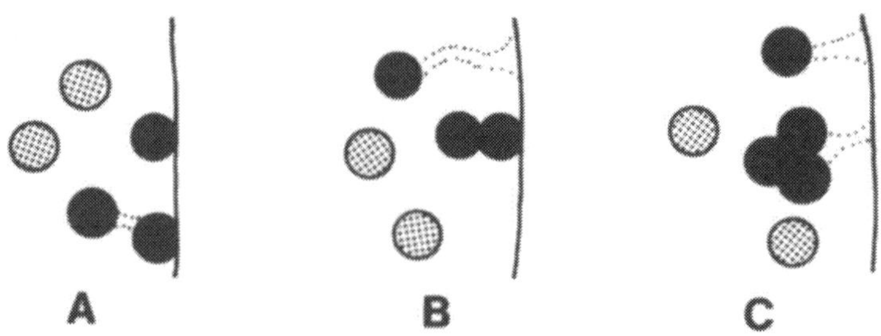
A
B
C

Fig. 13. The titer of hemolymph insulin-like peptide (ILP) in slow, medium, and fast growing snails (*Helisoma*)

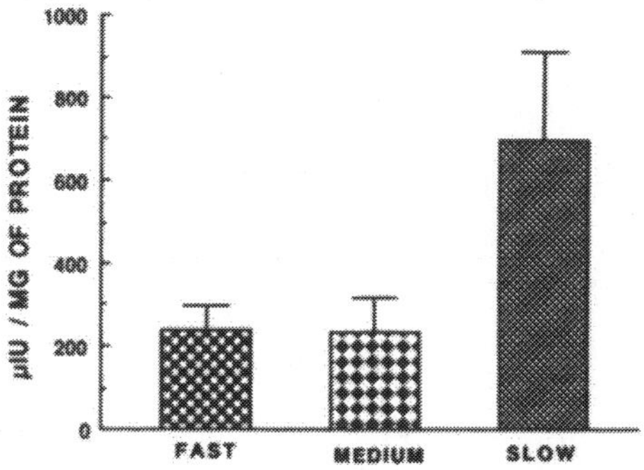

snails maintained in total darkness (00L:24D) grew significantly faster than those in 24L:00D and 12L:12D photoperiods [2, 11].

Using porcine and/or human insulin antiserum, we showed that the MDC as well as the median lip nerve in *Helisoma* contained insulin-like immunoreactive material (Fig. 2). Serial sectioning of the brain showed that not all cells were immunoreactive nor were they uniformly stained. Similar observations have been made on other pulmonates [29]. This may be partly explained in that bovine anti-insulin may not recognize the multiple forms of insulin-like material which are probably present in the population of MDC.

Detection of Insulin-like Peptide (ILP) by Radioimmunoassay (RIA) and Western Blotting in *H. Duryi*

The amount of ILP in the hemolymph of fast, medium, and slow growing snails was measured by RIA against human insulin. Surprisingly, the titer was highest in the slow growing snails (Fig. 13). This can be explained by the fact that the rate of use of ILP might be lowest in slow growing as opposed to fast growing snails, and thus its accumulation in the hemolymph in the former group. This

←

Fig. 8. *Helisoma*. Cross-section of median lip nerve showing the MDC axon terminals at their neurohemal area. *BM*, Basement membrane. *Bar* = 1.0 µm

Fig. 9. *Helisoma*. TARI-treated median lip nerve of a virgin snail in low magnification showing numerous electron dense exocytotically released granules. Note that the unreleased granules are less electron dense. *Bar* = 1.0 µm

Fig. 10. *Helisoma*. Higher magnification showing the excytotic release profile of a single granule. Note that the unreleased granules are less electron dense. *Bar* = 0.2 µm

Fig. 11. *Helisoma*. Release profile of multiple exocytosis. *Bar* = 0.2 µm

Fig. 12. *Helisoma*. Explanation of TARI-revealed exocytosis. **A** Exocytosis of single granules with or without attached plasma membrane. **B** Exocytosis of fused granules. **C** Exocytosis of many granules without attached plasma membrane. *Dotted line* represents membrane invagination or membrane caveolae above or below the plane of sections. Not to scale

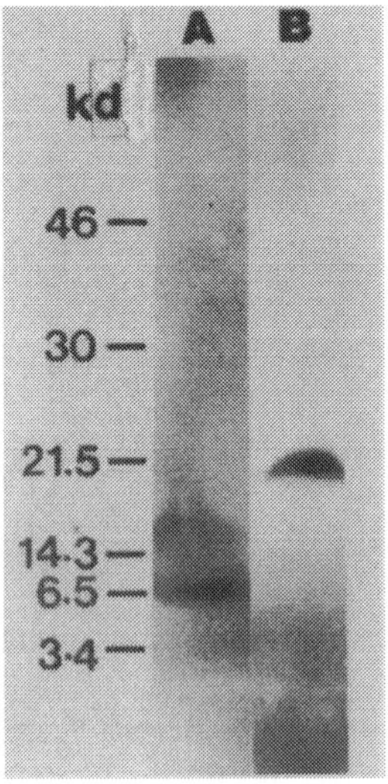

Fig. 14. Immunoblot of ILP in the hemolymph of *Helisoma duryi* (lane **A**) and *Otala lactea* (lane **B**). The samples were resolved by SDS-PAGE and then immunoblotted with bovine insulin antiserum. Note: two bands of 16 kd and 7 kd in *Helisoma* and one band of 21 kd in *Otala*

explanation relies on the premise that the rate of synthesis and release of ILP in slow and fast growing snails are identical. An alternative explanation would be that, at a higher concentration, ILP inhibits shell growth. Both high K^+ and 4- aminopyridine (4-AP) caused an increased release of ILP from MDC (H.R. Khan and A.S.M. Saleuddin, unpublished results). This release was maximum within 4 h after the addition of high K^+ and 4-AP. Similar results have been obtained with *L. stagnalis* [25]. After eletrophoresis in SDS PAGE followed by transfer to nitrocelluose, an immunoblotting technique revealed two insulin immunoreactive bands in the hemolymph of *Helisoma* but only one band in that of *Otala*. The peptides appeared to be 16 and 7 kd in *Helisoma* and 21 kd in *Otala lactea* (Fig. 14).

Fig. 15. Scanning micrograph of the mantle edge of *Helisoma* in vitro culture for 48 h. Note the newly formed periostracum (*P*) is protruding from the periostracel groove (*PG*). *VS*, Ventral surface; *MEG*, mantle edge gland. *Bar* = 100 µm

Fig. 16. Diagrammatic longitudinal section of the mantle edge of *Helisoma* showing the general histological features. Periostracum formation is initiated by periostracal cells (*PC*) but is thickened by the mantle edge gland (*MEG*). The shell part (*SH*) is formed by the dorsal epithelium (*DE*). The presumed sites of action of insulin-like peptides are shown by *lines*. *C*, Ciliary tufts; *CB*, receptor cell body; *CE*, ciliated epithelium; *DR*, dorsal receptors; *IN*, intrinsic neuron; *IND*, dorsal plexus intrinsic neuron; *MC*, mucous cell; *NV*, longitudinal branch of the pallial nerve; *PG*, periostracal groove; *RE*, receptor ending; *RP*, receptor process; *T*, tip of ventral mantle edge; *TZ*, transitional zone; *VS*, ventral surface. Not drawn to scale. Modified from [42]

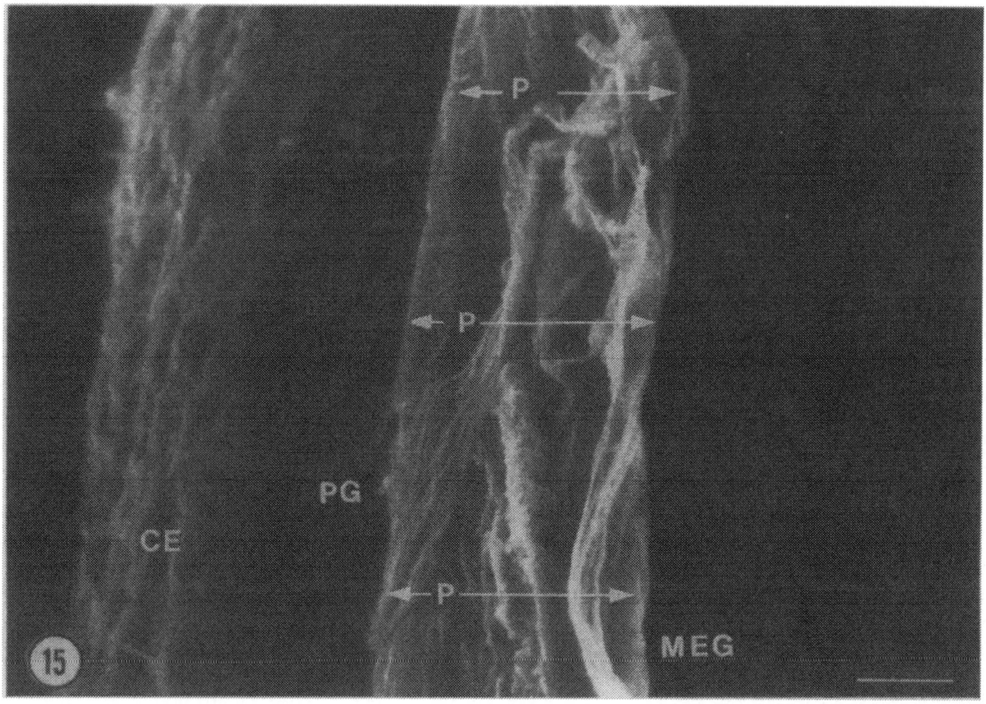
P
P
P
PG
CE
MEG
15

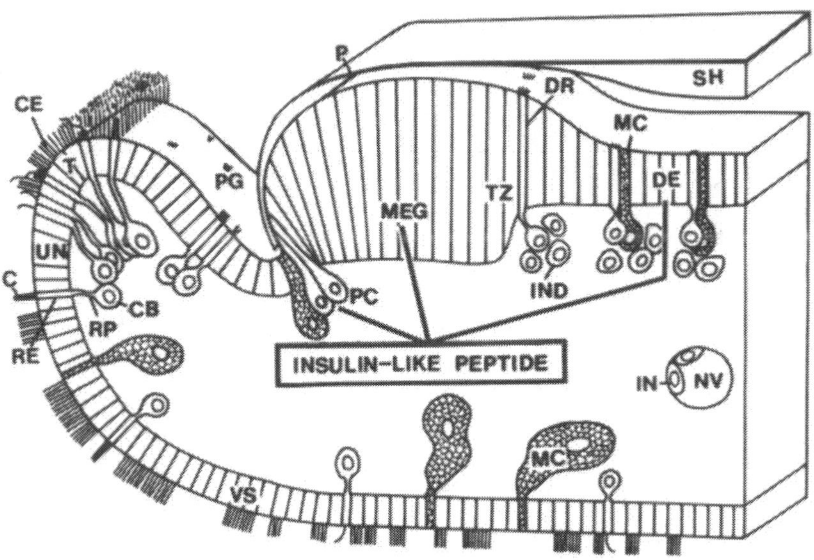
P
SH
CE
DR
MC
T
PG
DE
MEG
TZ
UN
C
PC
IND
CB
RP
RE
INSULIN-LIKE PEPTIDE
IN
NV
MC
VS

16

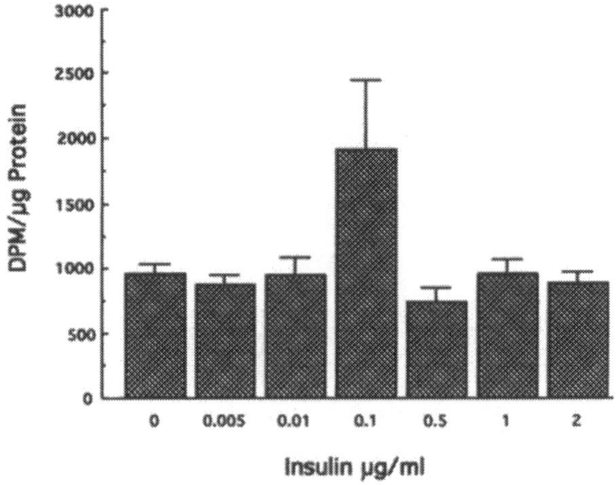

Fig. 17. Dose-response curve of porcine insulin on protein synthesis in the mantle of *Helisoma* in vitro

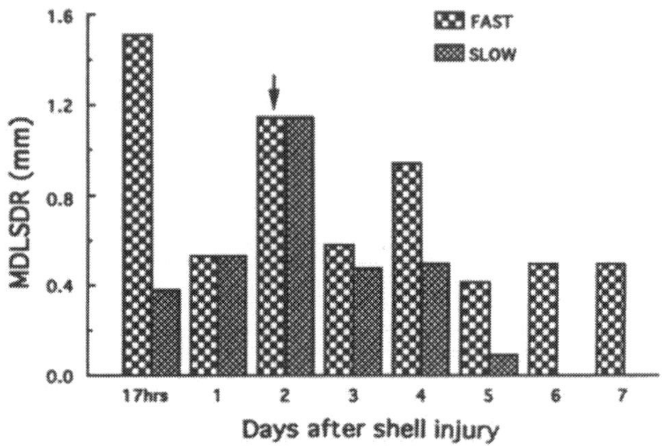

Fig. 18. Shell deposition rate (*MDLSDR*) in fast and slow growing snails (*Helisoma*) following shell injury at the edge. *Arrow* indicates the completion of shell repair

Effect of Porcine Insulin on Shell Growth and Shell Regeneration in *H. Duryi*

When a microgram of porcine insulin was injected into snails, the shell growth was stimulated. The significance of this observation is enhanced by the finding that the injection of porcine insulin antiserum inhibited shell growth [26]. One of the targets of ILP was the mantle which secretes the shell. When cultured in vitro, the mantle secreted the periostracum (Figs. 15, 16), and thus the effect of porcine insulin on periostracum synthesis could be assayed by measuring

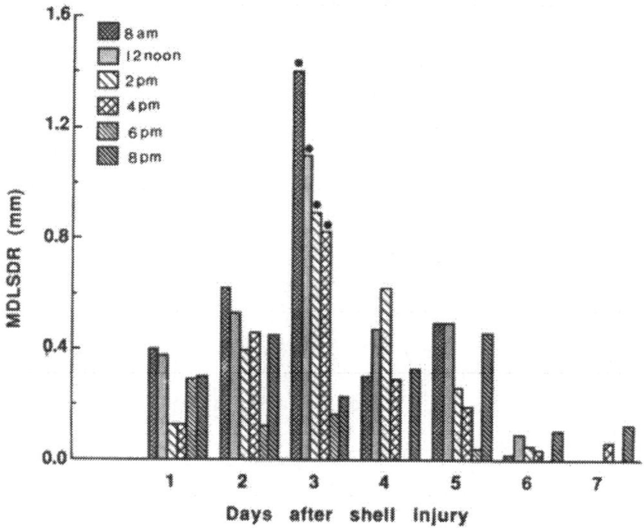

Fig. 19. Shell deposition rate (*MDLSDR*) in *Helisoma* following shell injury at the edge performed at different times of the day. *Solid circles* indicate resumption of shell growth at the edge

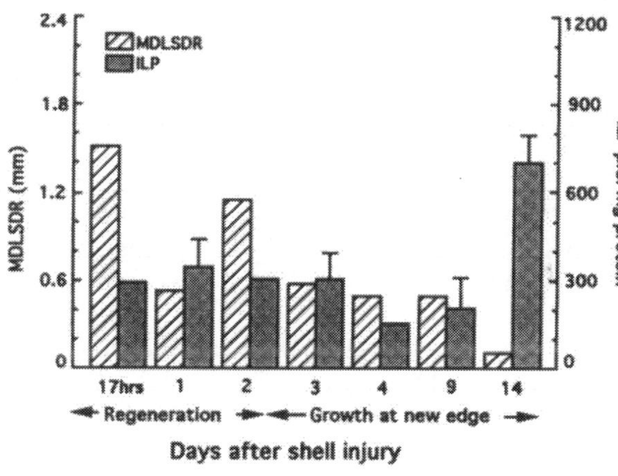

Fig. 20. Fluctuations in the hemolymph titers of ILP following shell injury at the edge in fast growing snails (*Helisoma*)

incorporation of tritiated amino acids into the periostracum in vitro [16, 30]. The mantle from a slow growing snail incorporated a significantly higher amount of amino acids in vitro when porcine insulin was present in the medium (Fig. 17). Similar results were also obtained with insulin-like growth factor II (IGF-II) (results not shown here).

In *Helisoma*, when a piece of shell was removed from the edge of the shell,

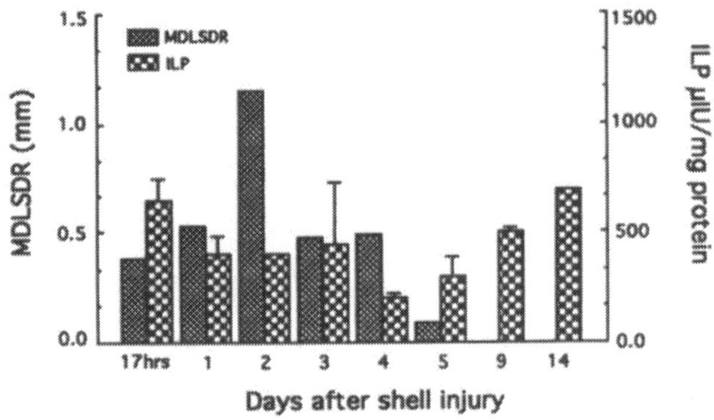

Fig. 21. Fluctuations in the hemolymph titers of ILP following shell injury at the edge in slow growing snails (*Helisoma*)

a new piece rapidly formed [2, 6], making this system suitable for designing experimental protocols for studying the hormonal titer during shell regeneration which involves rapid transport of organic and inorganic components. Normally, in fast growing snails, an injured area of 5 × 5 mm is repaired within 2 days, whereas in slow growing snails, it took nearly 5 days to complete the same repair process. (Fig. 18). In order to determine the best time to obtain the fastest regeneration, we damaged an uniform area of the shell every 2 hours, from 8 AM to 8 PM. Shell regeneration was best when injury was made between 8 AM and 4 PM. At 6 PM as well as at 8 PM, shell regeneration was slow (Fig. 19). Therefore, throughout the shell regeneration study, shells were damaged between 8 AM and 4 PM. The reason for this difference may become apparent when we will later discuss the ILP titer. We also found that the shell regeneration rate did fluctuate. Typically, in fast growing snails, the rate was high in the first 24 h, and the damaged area was fully repaired by day 2. Normal growth at the edge resumed by day 3 and continued at a steady rate until day 9 (Fig. 20). In slow growing snails, the duration of shell repair of the damaged area was longer, 5 days instead of 2 (Fig. 21). The peak rate of repair rate occurred at day 2, followed by a reduced rate until day 9, when there was no shell growth at all. Shell growth at the edge resumed at day 14 (results not shown). Having found differences in shell growth rates during shell regeneration in fast and slow growing snails, we wanted to see if ILP titers fluctuate during shell regeneration and, if they did indeed fluctuate, how did they correlate to shell repair rates in fast and slow growing snails? Figure 20 shows the ILP titer pattern during shell regeneration in fast growing snails over a 14-day period. During the first 24 h, when the repair rate was maximal, the ILP titer was low. At day 14, when the shell growth was low, the ILP titer was high. In slow growing snails, the general pattern of titer was similar to fast growing ones, in that the titer was high when the shell repair rate was slow and vice versa. Again, when the animals stopped growing, the titers were high (Fig. 21). It is worth emphasizing that ILP titers were higher in slow growing than in

Fig. 22. The effect of bovine insulin on growth (weight in grams) in young *Helix aspersa maxima* during 6 weeks of study: 0.5 µg insulin/gram weight was dissolved in 10 µl 1% bovine serum albumin in saline. *Arrows* indicate when injections of insulin were made (Courtesy of A. Gomot and L. Gomot)

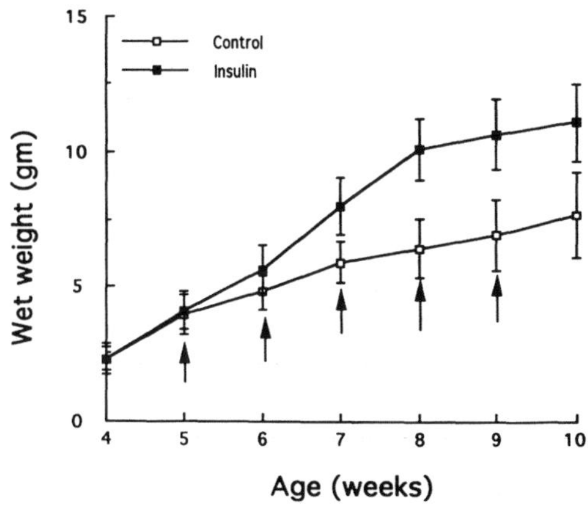

fast growing snails. The results we obtained from shell regeneration studies seem to correspond well with shell growth as far as the relationships of ILP titer of shell growth is concerned.

Is ILP Titer Controlled by Photoperiod in *H. Duryi*?

To study this, we first determined ILP titers every 2 hours over a 12L:12D period. Two major peaks were detected, one at 8:30 AM which followed the scoto phase, and the other at 6 PM immediately prior to the lights-off time (7:30 PM). The titer remained low during the scoto phase, with a small peak at 2 AM. When the snails were acclimatized in 24D:OL, two peaks were still present, but they had shifted by about 2 h, occurring at 6 AM and 4 PM. These peaks were also present when snails were raised in 24L:OOD, but the ILP titers decreased significantly. These experiments do not prove that ILP titer follows a circadian rhythm but rather suggest the possibility. We are now conducting further experiments to demonstrate conclusively the presence of a rhythm in ILP titer and that it is influenced by photoperiod. Such rhythm in hormone titers occurs in vertebrates and insects [31, 32]. As far as we are aware, this is the first report of the presence of a rhythm of ILP in molluscs. ILP could be involved in shell growth and/or shell regeneration by acting in concert on three main cellular components, the periostracal cells, mantle edge gland, and the dorsal epithelium, on the mantle edge (Fig. 16).

ILP in Other Molluscs

We have already mentioned that LGC and canopy cells in *L. stagnalis* transcribe molluscan insulin-like peptides whose molecular structure has been determined by recombinant DNA technology [24]. Previous to this finding, the presence of insulin-like material in molluscs was suggested by Thorpe and Duve [33]. By immunohistochemical techniques using anti-MIP-C, immunoreactive cells have been localized in neurons in the CNS of several molluscs [21]. In *Helix aspersa*, immunoreactive cells were detected in the cerebral ganglia and, in *Limax maximus*, similar cells were found in the pedal and pleural ganglia. In *Aplysia californica*, immunoreactive cells were found in all ganglia of the CNS [21]. Using anti-porcine insulin, we have localized immunoreactive cells in all ganglia of the CNS of *Otala lactea*; however, many were located in the mesocerebrum (H.R. Khan and A.S.M. Saleuddin, unpublished results). ILP has also been found in the hemolymph, digestive gland (hepatopancreas), and in brain extracts of *Otala lactea* and *H. aspersa*. The amounts of ILP in the hemolymph and digestive gland are 90 ± 20 uIU/mg protein and 1000 ± 50 uIU/mg protein, respectively, in *H. aspersa*, and 413 ± 97 uIU/mg protein and 2077 ± 56 uIU/mg protein, respectively, in *O. lactea*. Injections of microgram quantities of bovine insulin caused young *H. aspersa maxima* to grow significantly large compared to controls over a 6-week period (A. Gomot and L. Gomot, unpublished results) (Fig. 22). A few comments about the use of immunocytochemical data using mammalian insulin antiserum are necessary here. It is well accepted that positive reactions of cells/products to certain antibodies do not necessarily mean that the product(s) are identical to the structure of the antigen generating the probe. However, immunohistochemistry has been a popular technique to localize vertebrate-like peptides in invertebrates and thus a large volume of work has been published in recent years. In many of these publications, only the localization of vertebrate hormones in various tissues of invertebrates has been sought, without considering the possible functions of these peptides/hormones. Our rationale for localizing ILP is that a molecule of this type is involved in shell growth, as we have documented by in vitro and in vivo experiments in this study. We were at first surprised to find that porcine insulin and IGF-II stimulated the periostracum synthesis by the mantle of *Helisoma* in vitro and, furthermore, that the injection of anti-insulin into snails caused retardation in shell growth. The mammalian insulin receptor binding site involves residues 8, 9, and 10 in the A chain and residues 2, 3, and 4 in the B chain. These residues have proven to vary considerably in many invertebrate insulin-like peptides [25]. Of course it is possible that some other site on the porcine insulin molecule is responsible for binding to the *Helisoma* ILP receptor or that the *Helisoma* receptor binding site is by chance very similar to that of the vertebrate. It has been suggested that mammalian anti-insulin does not bind to molluscan insulin-like peptide (MIP) or to other invertebrate insulin-like peptides; however, we can detect ILP in the hemolymph, CNS, and digestive gland of *Helisoma*, *Helix*, and *Otala* with anti-human insulin ([26]). Perhaps the ILP in *Helisoma* is structurally more similar to mammalian insulin than is *Lymnaea* MIP. Alternatively, the receptors for ILP in *Helisoma* tissues have a wider receptivity than the more restrictive MIP receptors.

Our findings demonstrate the presence of an ILP in the neurosecretory MDC and that this peptide is involved in shell growth. We are now using recombinant DNA technology to elucidate the gene structure and expression of ILP in *Helisoma*.

Somatostatin-like Growth Factor in Pulmonate Molluscs

Schot et al. [23] reported that LGC in *L. Stagnalis* contained immunoreactive somatostatin-like material (SSLM) and, because these cells were implicated in the production of a growth hormone, SSLM was speculated to be the growth hormone in this snail. This report prompted Grimm-Jørgensen [34] to undertake a series of experiments to investigate the role of SSLM on shell growth in *L. stagnalis* and *Physa*. She found a significantly higher amount of SSLM in fast growing young snails than in slow growing older snails. By HPLC, it was shown that SSLM occured in the blood and that the SSLM originated from the CNS. Finally, by use of in vitro bioassay, she showed that a factor from the CNS stimulated the secretion of tyrosine-rich material by the mantle. However, when mammalian somatostatin was used, the synthesis and secretion of tyrosine-rich material was not altered. In *H. aspersa*, Marchand et al. [35] detected SSLM in the hemolymph, CNS, hepatopancreas (digestive gland), and mantle by immunocytochemistry and immunoassay. They observed fluctuations in the level of SSLM in the hemolymph and in other tissues following shell injury in young and adult snails. The following major points emerge from their studies. Except in the CNS, the level of SSLM is higher in young than old, shell regenerating snails. Regenerating snails from short days (8L:16D) have more SSLM than those kept in long days (18L:6D). In young, shell regenerating snails, the level of SSLM increases during the early phase (2–3 h) whereas it increases during the later part of shell regeneration (24–48 h) in adults. They also found that injections of nanogram quantities of mammalian somatostatin caused an increase not only in the volume of the periostracal gland but also in its secretory products in both adult and young snails. The SSLM was purified and partially sequenced and it was found to be quite different from mammalian somatostatin [36]. These studies [34–36] suggested that an SSLM is involved in shell formation and/or shell regeneration and thus it could be a growth hormone. Ebberink and Joosse [37] suggested that the growth hormone in *L. stagnalis* has a molecular weight of 1 kd and thus is in line with the molecular weight of mammalian somatostatin 14. They found the growth factor activities in the two median lip nerves which comprise the neurohemal area of LGC. Median lip nerves are also the sites from which DeWith et al. [38] isolated two natremic factors, each with a molecular weight of 4 kd, which are involved in Na^+ influx into the skin (mantle) of *L. stagnalis*. Joosse [39] suggested that the natremic factors are synthesized by LGC. Smit et al. [24] isolated and sequenced an insulin-like peptide from the peptides of the median lip nerves by using recombinant DNA techniques and termed it molluscan insulin-like peptide (MIP). In situ hybridization showed that LGC and canopy cells of the lateral lobes produce this peptide. So far, we have seen that LGC in *L. stagnalis* possibly produce a growth hormone (a SSLM), two natremic factors, and an MIP. In *H. duryi*, the MDC are homologous to LGC of *L. stagnalis*, but we prefer to call them MDC because of their location in the cerebral ganglia and because they do not stain green with the alcian yellow and alcian blue technique. These cells seem to produce an immunoreactive ILP. Various lines of studies in our laboratory indicated that this peptide acts as a shell growth hormone and/or perhaps as a general growth hormone. Both in vivo and in vitro experiments showed that mammalian insulin and insulin-like growth factors (IGF) stimulate periostracum synthesis as well as shell growth.

While the presence of MIP in LGC in *L. stagnalis* is conclusive, the occurrence of natremic factors and SSLM in these cells is not. Co-localization of these peptides in the small cell has not been shown. However, it is likely that among LGC, 3 distinct populations exist, each producing a category of peptide(s) mentioned above. It is tempting to suggest that MIP and SSLM-like producing LGC cells occur in close association, and that SSLM cells control MIP cells. Similar close associations have been found between B cells and immunoreactive somatostatin-14 cell in the pancreas of salmonids [40]. We are now working on co-localization of SSLM-like and ILP in MDC in *Helisoma*. Additionally, we are studying the effect of mammalian somatostatin on periostracal protein synthesis in the mantle in vitro in order to determine whether or not its effects in growth are direct.

Growth Factor in Non-Pulmonate Molluscs

As already mentioned, Lubet [4] was the first to show that the cerebral ganglia of the prosobranch *Crepidula fornicata* contained a factor which stimulated body growth. Le Gall [41], while working on the same prosobranch, performed a series of extirpation and reimplanation experiments to show that certain cells in the anterior parts of the cerebral ganglia were involved in body growth. Extirpation of these cells caused cessation of body growth. This could be partly corrected by implantation of cerebral ganglia. Furthermore, he reported that severing of nerves arising from the cerebral ganglia also caused retardation of body growth. He therefore concluded that both hormonal and nervous stimuli were necessary for body growth in *Crepidula*. It should be noted that he did not study shell growth per se.

Conclusions

The use of antisera against well characterized vertebrate hormones allowed the recognition and localization of related peptides in pulmonate molluscs. ILPs appear to be involved in the control of growth (shell) and shell regeneration. In addition, there seems to be a circadian rhythm of ILP in *H. duryi*. Although ILPs appear to be involved in the regulation of shell/body growth, at least one other peptide (somatostatin-like), originating from the same group of cells as the ILPs, has substantial effect on growth/shell in *Lymnaea*. While comparable studies in non-pulmonate molluscs are lacking to date, there is evidence of the presence of a growth factor in the CNS.

Acknowledgements. This work was supported by NSERC of Canada. We thank Professor L. Gomot of université de Besançon, France, for allowing us to use unpublished results. Thanks are also due to Dr. B.G. Loughton for help in preparing the manuscript and to Ms M-L. Ashton for drawing Fig. 3.

References

1. Lutz RA, Rhoads DC (1980) Growth patterns within the mollusan shell. In: Rhoads DC, Lutz RA (eds) Skeletal growth of aquatic organisms. Plenum, New York, PP 203–254
2. Kunigelis SC, Saleuddin ASM (1978) Regulation of shell growth in the pulmonate gastropod *Helisoma duryi*. Can J Zool 56:1975–1980
3. Dillaman RM, Saleuddin ASM, Jones GM (1976) Neurosecretion and shell regeneration in *Helisoma duryi* (Mollusca: Pulmonata). Can J Zool 54:1771–1778
4. Lubet P (1971) Influence des ganglion cérébroides sur la croissance dc *Crepidula fornicata* Phil. (Mollusque Mésogastéropode). C R Acad Sci [III] 273:2309–2311
5. Wong V, Saleuddin ASM (1972) Fine structure of normal and regenerated shell of *Hetisoma duryi*. Can J Zool 50:1563–1568
6. Kunigelis SC, Saleuddin ASM (1983) Shell repair rates and carbonic anhydrase activity during shell repair in *Helisoma duryi* (Mollusca). Can J Zool 61:597–602
7. Saleuddin ASM, Wilbur KM (1969) Shell regeneration in *Helix pomatia*. Can J Zool 47:51–53
8. Saleuddin ASM, Chan W (1969) Shell regeneration in *Helix*: Shell matrix composition and crystal formation. Can J Zool 47:1107–1111
9. Saleuddin ASM (1970) Electron microscopic study of the mantle of normal and regenerating *Helix*. Can J Zool 48:409–416
10. Khan HR, Saleuddin ASM (1979) Osmotic regulation and osmotically induced changes in the neurosecretory cells of the pulmonate snail *Helisoma*. Can J Zool 57:1371–1383
11. Saleuddin ASM, Kunigelis SC, Khan HR, Jones GM (1980) Possible control mechanisms in mineralization in *Helisoma duryi* (Mollusca: Pulmonata). In: Omori M, Watabe N (eds). The mechanisms of biomineralization in animals and plants. Tokai University Press, Tokyo, pp 121–129
12. Subramanyan OV (1973) Neuroendocrine control of calcium levels in the blood of *Ariophanta semirugata*, a terrestrial pulmonate snail. Endocrinol Exp (Bratisl) 7:315–317
13. Burton RF (1977) Haemolymph calcium in Helix pomatia: Effects of EGTA, ganglion extracts, ecdysterone, cyclic AMP and ionophase A23187. Comp Biochem Physiol [C] 57:135–137
14. Geraerts WPM (1976a) Control of growth by the neurosecretory hormone of the light green cells in the freshwater snail *Lymnaea stagnalis*. Gen Comp Endocrinol 29:61–71

15. Geraerts WPM (1976b) The role of lateral lobes in the control of growth and reproduction in the hermaphrodite freshwater snail *Lymnaea stagnalis*. Gen Comp Endocrinol 29:97–108
16. Kunigelis SC, Saleuddin ASM (1984) Studies on the *in vitro* formation of periostracum in *Helisoma duryi*: The influence of the dorsal epithelium on calcium incorporation. Can J Zool 62:1177–1180
17. Saleuddin ASM, Mukai ST, Khan HR (1990) Hormonal control of reproduction and growth in the freshwater snail *Helisoma duryi* (Mollusca: Pulmonata). In: Loughton BG, Saleuddin ASM (eds) Neurobiology and endocrinology of selected invertebrates. Captus University Press, Toronto, pp 163–182
18. Dogterom AA, van Loenhout H, vander Schors RC (1979) The effect of growth hormone of *Lymnaea stagnalis* on shell formation. Gen Comp Endocrinol 39:63–68
19. Dogterom AA, Jentjens T (1980) The effect of growth hormone of the pond snail *Lumnaea stagnalis* on periostracum formation. Comp Biochem Physiol [A] 66: 687–690
20. Dogterom AA, Doderer A (1981) A hormone dependent calcium-binding protein in the mantle edge of the freshwater snail *Lymnaea stagnalis*. Calcif Tissue Res 33:505–508
21. Boer HH, van Minnen J, Kerkhoven RM (1991) Evolutionary aspects of morphology and function of peptidergic systems, with particular reference to the pond snail *Lymnaea* stagnalis. In: Florey E, Stefano GB (eds) Comparative aspects of neuropeptide function. Manchester University Press, Manchester, pp 154–171
22. Kunigelis SC, Saleuddin ASM (1985) Studies on the *in vitro* formation of periostracum in *Helisoma duryi*: The influence of the brain. J Comp Physiol 13:177–183
23. Schot LPC, Boer HH, Swaab DF, van Noorden S (1981) Immunocytochemical demonstration of peptidergic neurons in the central nervous system of the pond snail *Lymnaea stagnalis* with anti-sera raised to biologically active peptides of vertebrates. Cell Tissue Res 216:273–291
24. Smit AB, Vreugdenhil E, Ebberink RHM, Geraerts WPM, Klootwijk J, Joosse J (1988) Growth controlling molluscan neurons produce the precursor of an insulin-related peptide. Nature 331:535–538
25. Ebberink RHM, Smit AB, van Minnen J (1989) The insulin family: Evolution of structure and function in vertebrates and invertebrates. Biol Bull 177:176–182
26. Saleuddin ASM, Khan HR, Sevala M, Sevala V (1991) Hormonal control of confirmed shell growth in the snail *Helisoma duryi* (Mollusca: Gastropoda). In: Suga S, Nakahara H (eds) Mechanisms and phylogeny of mineralization in biological systems. Springer, Tokyo, pp 161–165
27. Benjamin PR, Swindale NV, Slade CT (1976) Electrophysiology of identified neurosecretory neurones in the pond snail *Lymnaea stagnalis* (L.). In: Salanki J (ed) Neurobiology of invertebrates. Akademiai Kiado, Budapest, pp 85–100
28. Vleiger TA de, Lodder JC (1983) Morphological and electrophysiological features of the growth producing light green cells of *Lymnaea stagnalis*. In: Lever J, Boer HH (eds) Molluscan neuroendocrinology. North-Holland, Amsterdam, pp 81–85
29. Minnen J van, Schallig H (1990) Demonstration of insulin-related substances in the central nervous system of pulmonates and *Aplysia californica*. Cell Tissue Res 260:381–386
30. Saleuddin ASM, Kunigelis SC (1984) Neuroendocrine control mechanisms in shell formation. Am Zool 24:911–916
31. Saunders DS (1979) Insect clocks. Pergamon, Oxford, pp 31–40

32. Menaker M, Binkley S (1981) Normal and endocrine control of circadian rhythms in the vertebrates. In: Aschoff J (ed) Handbook of behavioral neurobiology, vol. 4. Biological rhythms. Plenum, New York, pp 234–255
33. Thorpe A, Duve H (1984). Insulin and glucagon-like peptides in insects and molluscs. Mol Physiol 5:235–260
34. Grimm-Jørgensen Y (1983) Possible physiological roles of thyrotropin releasing hormone and a somatostatin-like peptide in gastropods. In: Lever J, Boer HH (eds) Molluscan neuroendocrinology. North-Holland, Amsterdam, pp 21–27
35. Marchand CR, Assaka L, Strosser MT (1989) Variations of somatostatin-like immunoreactivity in the circumesophageal ganglia, the hepatopancreas, the mantle edge and the hemolymph of shell-repairing snails (*Helix aspersa*). Gen Comp Endocrinol 73:59–68
36. Assaka L (1988) Etude biochimique d'une (ou de) substance(s) immunologiquement apparentée(s) ā la somatostatine-14 chez l'escargot *Helix aspersa*: Correlation avec la croissancé et lá réparation de la coquille. Thesis, Université de Besancon, France
37. Ebberink RHM, Joosse J (1985) Molecular properties of various snail peptides from brain and gut. Peptides 6 (Suppl):451–457
38. de With ND, Ebberink RHM, vander Schors RC (1987) Neurohormonal control of sodium metabolism in the pulmonate freshwater snail *Lumnaea stagnalis*. In: Boer HH, Geraerts WPM, Joosse J (eds) Neurobiology, molluscan models. North-Holland, Amsterdam, pp 357–359
39. Joosse J (1988) The hormones of molluscs. In: Laufer H, Downer RGH (eds) Endocrinology of selected invertebrate types. Alan R. Liss, New York, pp 89–140
40. Nozaki M, Miyata K, Oota Y, Gorbman A, Plisetskaya EM (1988) Different cellular distributions of two somatostatins in brain and pancreas of salmonids, and their associations with insulin and glucagon secreting cells. Gen Comp Endocrinol 69:267–280
41. Le Gall P (1980) Etude expérimentale de l'association en chaine et de son influence sur la croissance et la sexualité chez la crépidule *Crepidula fornicata*. Université de Caen, France, pp 1–251
42. Jones GM, Saleuddin ASM (1978) Cellular mechanisms of periostracum formation in *Physa* spp. (Mollusca: Pulmonata). Can J Zool 56:2299–2311

Phosphoproteins of Oyster (*Crassostrea virginica*) Shell Organic Matrix

A.P. WHEELER[1]

Summary. As in most biominerals, the oyster shell is a composite of inorganic crystals and an organic matrix. Its mineral is primarily calcitic calcium carbonate and its matrix is made up largely of highly acidic phosphoproteins. Among molluscs, this type of matrix may be confined to those which form the foliated microstructure. These phosphoproteins form a continuum of molecular sizes ranging from modest molecular weight soluble proteins to insoluble complexes. The fact that similar sequences or even peptides can be identified among the various size classes suggests that all the classes may have common precursors. The primary structure of the major reverse phase HPLC fraction is roughly analogous to the phosphoryns of tooth dentin, in that it contains domains enriched in aspartic acid, other domains enriched in phosphoserine, and a terminal hydrophobic domain. The proteins have high affinities for crystal nuclei and calcite crystals, but not calcium, at high ionic strength. From these activities, it is hypothesized that the insoluble complexes bind nuclei, initiating crystal or crystal layer growth, and that the soluble components adsorb to crystal surfaces and, thereby, both control crystal morphology and limit crystal growth.

Key words: Phosphoproteins—Matrix—Mineralization—Shell—Molluscs

Introduction

The calcium carbonate shells of molluscs are similar to most biominerals in that they contain a small percentage of organic matrix by weight. One of the main functions of the matrix is to produce a composite with structural characteristics far more resistant to stress than that of inorganic carbonate. Further, as is generally postulated for biominerals, it seems evident that the matrix serves to regulate the growth of the mineral phase [1], producing structures clearly different from those grown inorganically.

[1] Department of Biological Sciences, Clemson University, Clemson, SC 29634, USA

In many ways the molluscan shell appears ideal for studying the role of matrix in mineral formation. The biomineral is plentiful and readily prepared for study. Furthermore, the process of matrix mineralization occurs in an extracellular compartment (the extrapallial fluid) defined by the forming shell on one side and a tissue layer (outer mantle epithelium) on the other. Therefore, the mineralization environment itself is theoretically accessible.

One of the limitations of using the molluscan shell system as a model for mineralization is the diversity of shell microstructures produced by this group. Some structures have been studied more than others; for example, microscopic observation of mature and growing nacre abound and some data on the matrix of this structure have been accumulated [2]. However, even the studies of nacre is scattered over a wide number of species.

To partially avoid the dilemma of trying to provide an appropriately balanced treatment of the findings from so many systems, studies on the matrix from a single species, the oyster *Crassostrea virginica*, will be highlighted and used as a point of departure for discussing results obtained for other molluscan matrix or matrix analog systems. The oyster is perhaps no better a model than any other molluscan system and no doubt has many properties that are not general to other molluscan systems. In particular, the presence of significant quantities of highly phosphorylated proteins may be, among the bivalves, unique to the foliated microstructure which makes up the majority of the oyster shell [3]. Interestingly, these matrix proteins may be analogous to similar proteins found in calcium phosphate systems, especially the phosphophoryns of tooth dentin [4]. However, it should be noted that the oyster, in general, is a well-studied species. Further, there has recently been an accumulation of data on the oyster matrix in which the author has been actively involved. It is hoped that this contribution will serve to draw attention to this interesting class of proteins.

Shell and Matrix Structure

Typical of many molluscs, the oyster shell is constructed principally of two layers which are microstructurally distinct. These are a thin outer prismatic layer and an inner foliated (calcitostracum; pseudonacre) layer which makes up the majority of the shell. Much like nacre, the foliated microstructure is laminar, with each sheet (laminae; lamellae) of the layer being made up of flattened polygonal crystal tablets (folia) which grow together. In both cases, the appropriate vertical sections through several lamellae can yield a "brick wall" appearance (Fig. 2) and the *c*-axis of the individual crystals is found more or less along the thickness of the crystals and approximately perpendicular to the surfaces of the shell [5]. In contrast to nacre, the crystals of the foliated structure are more elongated and may be irregular and are composed of calcite rather than aragonite. In addition, the manner in which the crystals grow and the final faces expressed is no doubt different for the two microstructures.

Several classes of matrix have been identified in molluscan shell [5]. They include fibrous sheets which may be reticulated [6] and which lie between layers of crystals (interlamellar) and at the edges of crystals (intercrystalline) (Fig. 2). In addition, the crystals themselves may be divided into blocks by matrix [5]. In some systems, the matrix is layered—especially the interlamellar matrix—with the innermost (crystal-facing) layer of the interlamellar and intercrystallic matrix having been referred to by some as the envelope [7]. In the oyster, the interlamellar matrix appears poorly developed and the intercrystalline matrix is quite thin. However, intracystalline sheets are evident, dividing the crystals vertically [5].

In addition to the morphological designations of matrix, a class of matrix has been identified based on solubility when shell is dissolved with dilute acids or, more commonly, near-neutral solutions of EDTA [8]. Although, as expected, much of the matrix is often insoluble upon dissolution of the shell (insoluble matrix; IM), soluble proteins and glyoproteins have been released upon shell dissolution as well in every case to date. The soluble matrix (SM) proteins are often highly anionic, deriving their charge from the carboxylate groups of aspartic and, to a lesser extent, glutamic acid, phosphoserine, and sulfated carbohydrate residues [1]. In general, it is thought that these soluble proteins play a significant role in crystal growth regulation in situ, as these proteins are capable of affecting mineral growth in vitro (see below). One should bear in mind that, at this stage, no single biological function has been definitely assigned to any given matrix protein.

The origin of the SM is in doubt, but it may derive from envelope layers of the intercrystalline and interlamellar matrix. This conclusion comes from the fact that certain species appear to have only this layer [7] and that the matrix from these species is almost entirely soluble [9]. In addition, dissolution of pieces of prismatic shell by acid can ostensibly result in isolated envelopes which, when analyzed, contain highly anionic proteins [10]. However, other origins of SM have been mentioned, including the reticulum of the sheets [11] and the intracrystalline matrix. Evidence for the latter can be inferred from microscopic observations of dissolved or partially dissolved sections of shell, including that of the oyster [5], and from the observation that, compared to the insoluble matrix, the soluble fraction appears to be selectively protected from oxidation when powdered shell is exposed to oxidant [8].

Isolation and Fractionation of Oyster Phosphoproteins

As has been described for numerous other calcium biominerals, the primary mode of extraction of matrix proteins involves dissolution of mineral with EDTA at near neutral pH. A method for rapid dissolution has been developed [12] in which the shell is dissolved in dialysis bags that are rotated through large volumes of EDTA. The rotation serves to keep the powdered shell suspended within the bag, effecting a reduction in the time for dissolution. To date, no protease inhibitors or denaturants have been added to the dissolution media, as is common practice for extracting proteins from vertebrate calcium biomineral such as teeth or bones. In these latter cases, the opportunity for tissue contamination is much greater than for the shell, which during growth comes to lie entirely outside of the mineral-forming tissue. Accordingly, Weiner [13] reported that gel electrophoretic patterns of *Mercenaria* were unaffected by the inclusion of such agents in the extraction media. In contrast, Keith et al. [14] have shown different electrophoretic patterns of *Mytilus* shell extracts depending on whether or not protease inhibitors plus denaturing agents are included in the extraction media. The question of whether or not entities identified by electrophoresis constitute a significant fraction of the total protein extracted is not known (see below), making interpretation of these findings difficult.

As is normally the procedure for matrix extracts, the oyster protein is first fractionated into soluble and insoluble fractions, with the latter classified as that which forms into pellets upon centrifugation. The pellet from oyster has two distinct layers. The upper layer is a viscous, translucent material which is referred to as light insoluble matrix (LIM). The lower layer is dark and relatively particulate and has been termed dark insoluble matrix (DIM). The soluble matrix from the supernatant has been frationated by gel chromatography on Sephacryl 300, which resolves proteins having two classes of molecular weights, those greater than about 500 kd (SE I) and those with a mean around 50 kd (SE II). Using other gel chromatographic media, it has been determined that SE I can be further sub-fractionated with components having molecular weights ranging into the millions. In general, it appears that the soluble proteins form a continuum of molecular weights with a possible discrete class represented by SE II.

The soluble fractions can also be fractionated by ion exchange chromatography or reverse-phase HPLC. In DEAE-ion exchange, a pattern of elution for oyster SM [15] is roughly equivalent to that described for other molluscan species [9]. This technique has not proven especially useful in achieving better definition of fractions of matrix, in that large molecular weight fractions elute at zero ionic strength, probably because they are so large that only a small percentage of their ionic groups have access to the resin. Further, much, if not all, of SE II elutes near 0.3 M ionic strength with no distinct evidence of fractionation. This lack of fractionation is even more pronounced when various discrete reverse-phase HPLC fractions elute at the same ionic strength in step gradients (A.P. Wheeler, unpublished). However, more refined continuous gradients may result in subfractionation of SM.

Reverse-phase HPLC has resolved three fractions of oyster SE II termed RP-1, -2, and -3, with the first being the most hydrophilic [16]. Although RP-1 is typically collected as a near-void peak, attempts to further fractionate the peak led to the conclusion that it contains a discrete class of proteins [17]. When RP-2 was again subjected to chromatographic conditions identical to those used in the original fractionation, it partially separated into RP-1 and RP-3. This raises the question of whether or not there exists a relatively stable interaction between these latter two classes of proteins.

Wourms and Weiner [18] have also reported that two or three general reverse-phase classes of proteins form several species of molluscs, including a member of the same genus (*C. irredescens*). These fractions were obtained using a reverse-phase cartridge and step-wise elution. As with *C. virginica*, much of the protein elutes with the weakest buffer solvent and a second class elutes when the eluent is switched to a much stronger solvent (50% acetonitrile). A third fraction elutes with a step of dimethyl sulfoxide. Although it can be argued that these fractions roughly correspond to RP-1 through RP-3, the different solvent systems used in the two cases and possibly the different species studied make such a comparison tenuous at best (see below).

For several molluscan species, Weiner and co-workers have resolved numerous peaks from ion-exchange or reverse-phase cartridge fractions of soluble matrix using reverse-phase HPLC [13, 18, 19]. Since much of the total protein fractionationated in these studies appears at or near the onset of the gradient, it is difficult to assess the significance of the much smaller peaks which eluted later. Such a multitude of peaks was not observed for *C. virginica* when various gradients were used (K.W. Rusenko and A.P. Wheeler, unpublished data). However, in most of these latter cases, the sensitivity of the detector was adjusted to the scale of the major peaks and thus minor peaks, if any, could have escaped detection.

It should be mentioned that electrophoretic determination of protein constituents of soluble matrix from oyster has been hampered by the difficulty in staining the protein. Much like dentin phosphoproteins [20], these proteins are not reactive to Coomassie blue staining following standard acrylamide gel procedures. A lack of discrete Coomassie bands has been observed for soluble matrix from the calcite layers of *C. irredescens* and *C. gigas* [21], implying that similar types of proteins may be found in all the species of this genus. Discrete molecular weight components for other species have been identified using electrophoresis [14, 21], but the percent of the actual protein identified or entering the gel may be low. For the same reasons, caution should be applied to avoid overemphasizing the observation that SE II or RP-I occasionally produce dimeric bands around 50 kd when gels are stained using siliver techniques (A.P. Wheeler, unpublished data).

Composition of Oyster Fractions

The first observation that can be made regarding the oyster fractions is the similarity of amino acid composition among the components of the molecular weight fractions SE I and SE II and LIM (Table 1). In these cases, the aspartic acid, glycine, and serine make up the majority of the amino acids. In addition to a similarity of amino acid composition, all three fractions are highly phosphorylated, and, from β-elimination studies with base, it is clear that it is the serine groups which are phosphorylated. In summary, these are highly anionic proteins and, in this regard, typify in the extreme the soluble proteins extracted from diverse mineralized systems. For example, reverse-phase fractions with as high as 50% aspartic acid have been isolated from *C. irredescens* [18], and the envelope fraction isolated from prismatic layers of various bivalves was

Table 1. Composition of oyster shell matrix fractions[a]

Analysis	*LIM*	*SE I*	*SE II*	*RP-1*	*RP-3*	*DIM*
Amino acid[b]						
Asp	258	288	302	315	216	88
Thr	9	13	12	5	5	18
Ser	201	216	221	283	135	102
Glu	36	45	46	42	85	37
Pro	30	30	27	9	56	59
Gly	334	285	292	300	222	228
Ala	14	18	16	9	49	246
Cys	6	5	4			0
Val	14	11	11	2	23	35
Met	3	4	3			4
Ile	5	7	5	1	23	35
Leu	12	13	9	1	38	29
Dopa	0	0	0			8
Tyr	24	24	25	17	27	44
Phe	4	6	4	1	15	35
Lys	21	16	11	6	18	8
Arg	22	14	5	3	22	18
C/HP[c]	4.3	4.3	5.1	15.9	1.7	0.35
Carbohydrate	35.0	33.5	14.4	14.4	46.4	
Phosphate:[d]						
Total	109	129	140	153	49	
By OD_{240}	101		132	141	16	

[a] *LIM* is the light (translucent) insoluble matrix which forms the upper part of the pellet of oyster shell EDTA extracts upon centrifugation. *DIM* is dark insoluble matrix which is particulate and sediments under LIM. *SE I* and *SE II* are the principal molecular weight classes of soluble matrix obtained on Sephacryl S-300, with I containing material greater than about 500 kd and II a broad peak whose molecular weight is about 50 kd. *RP-1* and *3* are reverse-phase HPLC fractions of SE II, with RP-1 being the more hydrophilic and approx. 54 kd and RP-3 the more hydrophobic and about 70 kd. Data from [15–17]

[b] Residues per thousand

[c] Sum of the residues per thousand for the charged amino acids (asp, glu, lys, arg) divided by the sum of hydrophobic amino acids (pro, ala, val, ile, leu, phe)

[d] Phosphate as calculated from dehydroalanine detected at 240 nm during incubation of fractions in NaOH

reported to contain in excess of 70% aspartic acid [10]. In addition, some matrices have been reported as having sulfated carbohydrates [8].

Although there are relatively few studies which compare IM to SM, in cases where the comparison is clearly made, the IM tends to be somewhat lower in anionic amino acids than SM and have higher relative concentrations of certain hydrophobic amino acids [22]. The universeality of this observation is unknown; however, in LIM from oyster, a clear case for a dichotomy between SM and IM based on composition would be hard to make. In contrast, DIM is clearly distinctive from the other matrix fractions, having much lower relative aspartate levels and much higher relative levels of hydrophobic amino acids. In particular, alanine is a major constituent of DIM and dihydroxyphenylalanine (dopa) is present in this fraction (Table 1). It is possible that similar IM fractions exist in other species and simply have not been separated. For example, Travis et al. [6] analyzed IM from oyster and found compositions that could be described as intermediate between DIM and LIM.

The insolubility of IM and, for that matter, the molecular weight continuum of the SM most likely result from cross-linking between molecular species. However, no cross-links have been definitively described to date. The presence of phenoloxidase in shell extracts or sections of various species, including a member of the genus *Crassostrea* [23, 24], would suggest the involvement of dopa residues in the cross-linking of DIM.

LIM can be solubilized and the molecular weight of SE I reduced upon treatment with 0.1 N NaOH. This treatment results in fragments that have chromatographic properties similar to SM when passed over gel permeation [15] or reverse-phase HPLC media [17]. Whether the base is disrupting cross-links or fragmenting the backbone of the molecules is unknown. However, a linkage that would be especially base susceptible is a phosphodiester cross-link, a bond which has been described in other systems [25]. Given the significant levels of phosphorylation of both SE I and LIM, it is possible that at least a small percentage of the phosphates are involved in such a linkage.

Other potential cross-links have not been explored, and therefore cannot be excluded. For example, histidinoalanine has been described in both the phosphoprotein particles of molluscan mantle and extrapallial fluid [26], as well as, with lysinoalanine, in dentin phosphophoryn [27]. A histidinoalanine type of linkage is made less likely by the fact that histidine is present in very low quantities in the oyster phosphoproteins; however, lysinoalanine is a distinct possibility. It is possible that non-covalent interactions stabilize LIM or SE I. However, treatment of SE I with denaturants of non-covalent linkages, such as sodium dodecyl sulfate or cetylpyridinium chloride, does not result in reduction in size of this fraction.

Compositions of the reverse-phase fractions RP-1 and RP-3 indicate that there is actually some heterogeneity within the matrix phosphoproteins (Table 1). However, one must bear in mind that the heterogeneity may result from a relatively small percentage of the total protein in that RP-1 makes up greater than 90% of the SE II fractionated by reverse-phase. This fact is evident from the similarity of compositon of SE II and RP-1.

One of the most significant differences between RP-1 and RP-3 fractions is the relative concentration of phosphate and carbohydrate, with RP-3 being lower in the former and higher in the latter. Further, the phosphate of RP-1 appears to be more or less entirely linked via serine residues, as the total phosphate of the fraction agrees with that which can be determined as dehydroalanine following β-elimination. In contrast, using the same criteria, only a fraction of the phosphate of RP-3 appears to be O-linked via serine, suggesting that other O-linkages or N-linkages are possible.

Additional differences between the fractions are reflected by the amino acid composition of the two samples in that Asp, Ser, and Gly make up approximately 90% of the amino acids of RP-1 and less than 60% of RP-3. Further, as expected from their relative elution from reverse-phase HPLC, the fraction of hydrophobic residues in RP-3 is much higher than in RP-1. In comparing the "A" or most hydrophilic fraction isolated by Wourms and Weiner [18] from *C. irredescens* with RP-1 of *C. viriginica*, a significant disparity is observed in that although the "A" fraction is also 90% Asp + Ser + Gly, it is less than 10% Ser. The "C" fraction of Wourms and Weiner roughly corresponds to RP-3 in the composition of amino acids and in the fact that they both may be more glycosylated than their more hydrophilic counterparts. However, it is also clear that these more hydrophobic fractions isolated by the two methods differ significantly in detail. In short, it is apparent that RP-1 and RP-3 are unique fractions, and the extent to which they may be present in other species, especially those of the same genus or those having the foliated microstructure, will require extraction and fractionation of the matrix from various species using the same procedures as employed for *C. virginica*.

It should be noted that reverse-phase fractions that elute much like RP-1 and RP-3 have been obtained from preparations of base-solubilized LIM [17]. Predictably, their composition differs from the soluble matrix fractions in that base-induced β-elimination at serine will cause a loss of that amino acid. However, taking that loss into account, the residual amino acid distribution remains remarkably similar to the fractions from SM. This finding accentuates the overall homogeneity of the phosphoprotein complexes.

Sulfate, presumably as sulfated carbohydrate, has been identified as a significant component of both *Mercenaria mercenaria* and *Mytilus californianus* SM proteins, being 10% by weight of the former [8] and in a ratio of about 70 residues per 100 amino acids in the latter [28]. Simkiss [29], using the results of infrared analyses, suggested the presence of sulfate in matrix extracts of *C. virginica* as well. Chemical analysis of SE II resulted in only approximately 1% sulfate by weight (D.M. Swift, unpublished data). It is interesting to note that neither *Mytilus edulis* nor *M. mercenaria* SM is phosphorylated to any extent [3]. Given that all these species have different shell microstructures, the dichotomy of sulfated versus phosphorylated matrix raises the question of whether the type of anion found in matrix is an important factor in controlling the shell crystal architecture.

Primary Structure of Phosphoproteins

Previous to studies of the oyster matrix, much of the evidence regarding the primary sequence of molluscs was interpreted as suggesting that these proteins contain domains of $(\text{Asp-X})_n$ in which X is primarily Gly or Ser. These conclusions were based on the finding that Asp, Ser, and Gly were released from SM proteins during a mild acid hydrolysis [13, 30–32]. Although this technique is selective for peptide bonds on either side of aspartic acid residues, they are not entirely specific for the release of this amino acid [1, 16, 17]. Consequently, if the hydrolysis is continued for extended periods, as it was in the above-mentioned studies, then the accumulated release of other amino acids is expected even if they are not adjacent to aspartic acid.

To avoid this complication, time-course hydrolyses were performed on a number of oyster phosphoprotein fractions including LIM, RP-1, and RP-3. In every case, the rate of release of aspartic acid far exceeded that of Gly or Ser [16, 17]. Further, Gly and Ser release were linear, whereas the rate of release of Asp approached a plateau in 2–4 h. These data, along with those for the hydrolysis of model peptides, suggest that Asp-Asp bonds are quite common in the proteins, possibly existing as domains of polyaspartate. If domains of Asp-X are present, they are certainly in the minority in oyster proteins. Further, domains of Asp-X-Y, if present at all, are also in the minority, as there was no evidence of Gly-Ser or Ser-Gly dipeptides released during the hydrolysis. These dipeptides were identified in the hydrolysates of SM from the calcitic-prismatic layer of *Pinctada margaritifera* [32]. However, they may be micro-structure-specific, as they were not identified in the nacreous layer of this species. It is interesting to note that the matrix of the calcitic-prismatic layer of *Mytilus californianus* contains the dipeptide Pro-Thr which is also specific to that layer [13].

Additional sequence information has been obtained for RP-1 using reverse-phase HPLC fractions of the residual material following mild acid hydrolysis [16, 17] and fragments produced by hydroxylamine cleavage [33]. In both cases, the fragments ranged dramatically in relative hydrophobicity, indicating distinct domains. One of the more hydrophilic fragments produced by acid hydrolysis, when subjected to automated Edman degradation, revealed an NH_2-terminal sequence composed almost entirely of serine. Because of the hydrophilicity of the fragment and the fact that phosphate is concentrated in the dialyzed residue over the native material, this fragment may represent a phosphoserine-enriched domain. Edman degradation of several of the hydroxylamine fragments revealed runs of Asp, confirming the findings from mild acid hydrolysis. Furthermore, in the fragments sequenced to date, $(Asp\text{-}X)_n$ appears to be virtually absent, confirming the results of mild acid hydrolysis.

The identification of hydrophobic fragments from RP-1 is interesting in that only a small percent of the protein is composed of hydrophobic amino acids. From hydrolysis studies using a variety of carboxypeptidases, it is clear that one of those domains is at the carboxy terminus of the intact protein [16, 17]. One of the polyaspartate domains is at the carboxy terminus as well. The only other molluscan matrix fractions for which this type of sequencing has been attempted is a blot-transferred electrophoretic fraction isolated from *Mytilus edulis* [14]. In this case, a hydrophobic domain was tentatively identified at the amino terminus.

The distribution of the glycine residues in oyster phosphoprotein is not yet clear. However, because glycine in dialyzed residues obtained from mild acid hydrolysis is in about the same molar proportion as in the native protein, it could be surmised that there are domains of polyglycine or at least domains highly enriched for the amino acid. These domains would provide a large

NH2-	**[Aspn]x** **[(P)Sern]x** **[(Gly(P)Ser)n]x?** **[Glyn]x?**	**-[Aspn]-Hydrophobic-Lys-Gly-Tyr-Tyr-COOH**

Fig. 1. Suggested amino acid domains of the oyster phosphoprotein, RP-1. The carboxy terminus was elucidated by the use of carboxypeptidases. The terminal four amino acid sequence is consistent with the results from hydrolysis of three enzymes. The hydrophobic domain appears to contain phe, leu, ile, val, and ala—all minor components of RP-1. The sequence of these amino acids within the domain is not known. The elucidation of an Asp_n domain near the carboxy terminus was elucidated with carboxypeptidase P. The existence of Asp_n domains in general follows from the rate of release of this amino acid during mild acid hydrolysis and from partial sequence analysis of hydroxylamine-cleaved fragments. The existence of $(P)Ser_n$ regions follows from compositional analysis of the residue of mild acid hydrolysis and from partial sequence of reverse phase fractions from these and hydroxylamine-cleaved RP-1. The possibility of $Gly(P)Ser_n$ is suggested but not confirmed by these studies. Gly_n domains are entirely speculative at this time and based on the composition of residues following hydrolysis. Sequence analysis and reverse-phase profiles of cleavage fragments also suggest that some mixed domains exist as well (not shown). Structures summarized from data in [16, 17, 33]

degree of rotational freedom to the protein. A summary of the putative domains for RP-1 identified to date is given in Fig. 1.

A comparison can be made between RP-1 and phosphophoryn, the aspartic acid-rich phosphoprotein from tooth dentin. This latter protein also appears to contain hydrophobic domains in the end regions of the molecule and both Asp and (P)Ser domains [34]. The difference is the presence of putative Asp-(P)Ser domains in the dentin protein which may be absent in RP-1. To some extent it appears that phosphoproteins from the two mineral systems may be structurally analogous. The structural analogy indicates that they may function in similar ways despite the fact that the minerals evolved along different lines and at different periods.

Functional Aspects of Phosphoproteins

If one accepts the hypothesis that matrix serves to regulate crystal growth, it then follows that the oyster phosphoproteins must be shown to interact with calcite, preferably in more or less specific ways. For example, it is now well established that immobilized matrix molecules can induce crystal growth from slightly supersaturated solutions. One technique used to demonstrate this phenomenon involves extraction of matrix in fixatives so that SM remains associated with IM [35]. Other approaches involve attaching SM proteins to

Table 2. Calcium, nuclei, and crystal association properties of oyster phosphoproteins and synthetic peptides[a]

Calcium binding:	
Low ionic strength	SE II—affinity K, 10^{+4}, 10^{+5} M^{-1}
	Capacity, 1-2 μmol mg^{-1}
	LIM, polyasp—binding occurs
500 mM NaCl	SE II, SE I, polyasp-binding negligible
$CaCO_3$ Nucleation—order of inhibitory activity on weight basis in 500 mM NaCl[b]	
$Asp_{20}PSer > (Asp(P)Ser)_{10} > RP\text{-}1 \cong Asp_{20} > RP\text{-}3$ $(AspSer)_{10} \cong Asp_{10} > (AspSerGly)_{10} \cong (Asp(P)SerGly)_{10}$	
$CaCO_3$ Seed crystal-binding and growth inhibition in 500 mM NaCl:	
Molar affinity:	
$RP\text{-}1 \gg Asp_{19} \cong (AspSer)_{10} \cong (AspSerGly)_{10} > (Asp(P)SerGly)_{10}$	
Molar capacity[c]	
$Asp_{19} > (Asp(P)SerGly)_{10} > (AspSer)_{10} > (AspSerGly)_{10} \cong RP\text{-}1$	
Max % crystal growth inhibition when crystal sties saturated for peptide[d]	
$RP\text{-}1 = Asp_{19} = 100\% > (Asp(P)SerGly)_{10} > (AspSer)_{10} > (AspSerGly)_{10}$	

[a] Data from [12, 41, 42] and A.P. Wheeler and A.S. Mount, unpublished results. *(P)Ser* designation means partial phosphorylation of Ser

[b] The relative capability to inhibit nucleation should reflect the affinity of peptides for ion clusters or nuclei. On a molar basis the inhibitory activity of RP-1 would exceed that of all peptides tested to date, with the possible exception of polyaspartates having terminal PSer residues

[c] The low molar capacity of RP-1 relative to its high affinity and capability for inducing 100% inhibition suggests that it can bind to more than one growth site

[d] RP-1 achieves 100% inhibition when sites are only 50% saturated, suggesting that attachment of RP-1 to one Asp_{19} site influences growth at adjacent sites. The maximal percent inhibition for other peptides is identical to the percent of Asp_{19} sites to which they can bind, suggesting that binding sites are equivalent to growth sites and more than one kind of growth site exists on crystals

polystyrene films [28] or gel beads [36]. The idea in each case is to simulate the relationship between the crystal active component (polyanionic SM) and the framework molecules (IM). An ideal matrix material that combines the anionic and insoluble characteristics is the LIM of oyster, a system that has not been studied in this regard.

The mechanism for the induction of mineral by insolubilized matrix is often explained in terms of calcium binding to the anionic substrate which results in the accumulation of layers of both cations and subsequently anions, and thus formation of crystal nuclei at that site [37]. This hypothesis is based on the observation that SM has a low affinity, high capacity calcium binding when in solution, especially at low ionic strength [1] (Table 2). Using soluble oyster phosphoproteins, Wheeler et al. [12] has demonstrated that most of the calcium-binding capacity of these proteins is lost at physiological ionic strength. Probably more significant to the theories of induction of crystal growth is the question to what extent the insoluble phosphoprotein LIM, which binds calcium at low ionic strength (A.P. Wheeler and A.S. Mount, unpublished data), retains its binding capacity at physiological strength. It should be mentioned that the effect of ionic strength on oyster soluble protein stands in contrast to observations on phosophophoryn in which significant calcium binding is retained at high ionic strength [38]. Because studies with model proteins suggest that repetitive carboxyl-phosphate groups may enhance calcium binding [39], it may be that the lack of Asp-PSer domains in oyster matrix but apparent presence in phosphophoryn may explain their different binding properties.

An alternate hypothesis to explain the effect of immobilized polyanions on induction of crystal growth involves the binding of nuclei rather than ions from metastable solutions [12]. This is a credible hypothesis in that soluble oyster phosphoproteins, on a molar basis, are as good as or better than other polyanions as inhibitors of calcite induction in bulk solutions [40] (Table 2) and thus probably have a high affinity for nuclei. In addition, unlike calcium-binding affinity, a high affinity for nuclei is retained at elevated ionic strength, as inhibitory assays are routinely run in 500 mM NaCl. Based on studies with model peptides, it is clear that this inhibitory activity is quite sensitive to primary structure. That is, peptides with runs of polyanions (devoid of spacer amino acids), much like those found in oyster phosphoproteins, are especially effective inhibitors. The addition of PSer termini to polyaspartate peptides dramatically increases their efficacy as inhibitors of calcite induction [41] (Table 2), indicating that phosphorylated domains other than Asp-(P)Ser calcium-binding domains may be excellent nuclei-binding agents.

The oyster soluble phosphoproteins are also excellent inhibitors of calcite crystal growth and effect this inhibition through high affinity binding to crystal surfaces [42] (Table 2). For the natural proteins, the runs of polyaspartate no doubt contribute to the overall efficacy of RP-1 inhibition in that synthetic polyaspartate is more effective as a crystal growth inhibitor on a mole basis than $(\text{Ser-Asp})_n$ or $(\text{Gly-Ser-Asp})_n$ peptides. The addition of phosphate to Ser in these model peptides increases their efficacy as an inhibitor as well, suggesting an importance for the PSer of oyster matrix in its interaction with growing calcitic crystals. This was confirmed in studies in which alkaline phosphatase dephosphorylated SE II fractions were shown to have significantly lower inhibitory activity than the native protein. In fact, the activity approached that of non-phosphorylated SM from other species [3].

Most data on crystal growth inhibition have been amassed from in vitro systems. However, it has been established that these oyster phosphoproteins can inhibit sea urchin embryo calcitic spicule growth in vivo [43]. Furthermore, it was determined that the protein actually bound to the crystals, a process that involved transport of the protein through cellular layers. This transport would not necessarily be expected of polyanions. Possibly one function of the hydrophobic tail of the phosphoproteins is to allow for recognition by and transport across the cells.

The relevance of inhibition by matrix proteins measured in vitro to crystal growth in situ is not known. At first glance, this activity may be related to simple limitation of crystal growth either in preformed compartments [7] or by periodic secretion of relevant layers [44]. Another role for inhibitory matrix proteins might involve control of crystal microstructure. In this process, both immobilized and soluble matrix components might participate. For example, it has been shown that *Mytilus* SM proteins adsorbed onto a surface can increase the frequency of development of unusual crystal faces and growth along the *c*-axis of the crystal [45]. As mentioned earlier, this axis is elaborated in many molluscan microstructures, including the oyster foliated system. Whether immobilized oyster SM or LIM can induce similar effects is not known. However, given that cooperation between the sulfate and carboxylate groups of the *Mytilus* proteins may be necessary for growth regulation [28], a similar effect may be produced by the phosphate and carboxylate groups of oyster matrix.

Interactions of matrix from solution with growing crystals may effect growth modifications as well, especially if specific matrix proteins adsorb to and inhibit growth at specific sites or faces, forcing growth in other areas. For example, it has been shown that the maximal percent inhibition of crystal growth induced by Asp-containing peptides and the number of binding sites on calcite for the peptides varies with their primary structure [42] (Table 2). Specifically, Asp_{19} effects 100% maximal inhibition and binds twice the number of sites as $(Ser\text{-}Asp)_{10}$, which effects only about 50% maximal inhibition. In this regard, RP-1 appears to be rather non-specific in that, like polyaspartate, it effects 100% inhibition of crystal growth but has a molar affinity for the crystal surface much greater than even polyaspartate. Because, on a molar basis, RP-1 binds to fewer sites than does Asp_{19}, its higher affinity for crystals may indicate that the polymer is binding to more than one Asp_{19}-type site. In fact, it has been shown via scanning tunneling microscopy that RP-1 may form a ring structure when adsorbed onto the surface of calcite [33]. The significance of this configuration is unclear; but, if these proteins do absorb in this fashion, it suggests that they may have the capacity to bind to multiple sites.

The significance of the above-mentioned observations regarding the relationship of RP-1 to the regulation of mineral growth is not known. However, the non-specific effect of such strongly anionic structures seems to carry over to the morphology of the crystals grown in their presence. This applies to both oyster phosphoproteins [46] and to sulfated SM proteins from *Mytilus* [47]. In both these cases, gross changes in crystal morphology occur but no faces atypical of inorganically grown crystals appear. In contrast, matrix proteins of lower anion content can induce specific morphologies and also be more effectively incorporated into crystals than are the highly anionic proteins [47]. In this regard, it might be speculated that RP-3 (or similar peptides), having a much lower anion content, might regulate crystal morphology and subsequently become part of the intracrystalline matrix. The role of the highly charged proteins (RP-1) may be relegated to that of limitation of crystal growth.

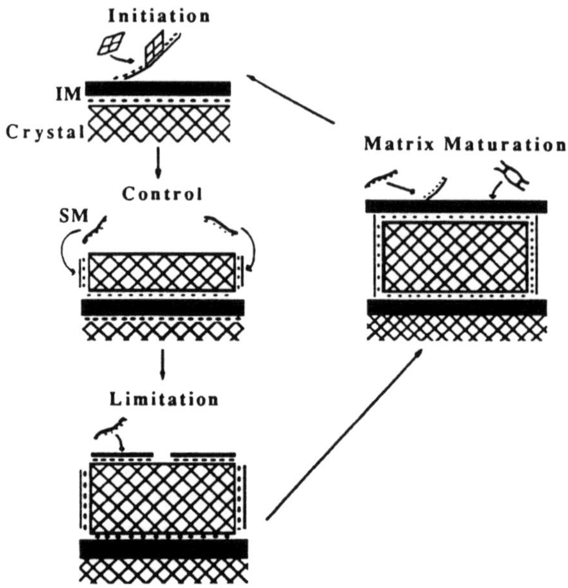

Fig. 2. A hypothetical scheme for the involvement of oyster matrix in crystal/crystal lamellae formation. *Initiation* The interlamellar matrix from the previous layer induces the next layer. Here, a polyanionic insoluble matrix such as LIM may be the induction factor. Nucleation is shown as binding nuclei rather than calcium since oyster matrix is a poor calcium binding agent at physiological ionic strength but apparently has a high affinity for nuclei under these conditions (Table 2). *Control* Soluble components may interact specifically with crystal sites, inhibiting growth in certain dimensions but allowing growth in others thus shaping the crystal. RP-3-type proteins, being less acidic, may have such selectivity and may be more easily overgrown, coming to reside as intracrystalline matrix. *Limitation* A highly charged, relatively non-specific inhibitor such as RP-1 may be involved in overall termination of growth. *Maturation* At this stage it is envisioned that structural elements (such as those cross-linked to form DIM) are added. In addition, the next layer of polyanion, which may include soluble elements, is cross-linked in some fashion to form a new LIM layer. Overall, a charged layer (envelope) comes to lie next to the crystal and other layers form between the envelopes. This process of matrix production is shown as being sequential with crystal growth. However, the matrix structure may be preformed, at least in part, and still function substantively as depicted

Conclusions

Despite their apparent complexity, the phosphoproteins that make up the majority of the oyster matrix may well consist of a fairly limited number of major classes of proteins. It is the interaction among these classes in time and space that no doubt controls the process of mineralization. A simple scheme of these processes is shown in Fig. 2. This scheme is not intended to be exhaustive or inclusive of all ideas or possibilities for matrix function. It does, however, reflect some of the organizing principles identified thus far. Furthermore, such a model indicates the need for continued study in several areas. For example, the interaction among the matrix components is far from clear. In addition to continued study on the interaction of components identified in extracts, knowing the form (or forms) of matrix molecules secreted by the mantle would help in the reconstruction of matrix molecular morphology. In turn, knowing how the matrix is structured comprehensively may help in the design of additional assays for its function. Continued analysis of the composition and primary sequences of matrix components is clearly a requisite to understanding both overall matrix structure and function. Combining these features of the oyster phosphoprotein-foliated calcite microstructure system would facilitate endeavors involving other shell structures in particular and mineralizing systems in general.

References

1. Wheeler AP, Sikes CS (1989) Matrix-crystal interactions in $CaCO_3$ biomineralization. In: Mann S, Webb J, Williams RJP (eds) Biomineralization: Chemical and biochemical perspectives. VCH, Weinheim, pp 95–131
2. Weiner S (1986) Organization of extracellularly mineralized tissues: A comparative study of biological crystal growth. CRC Crit Rev Biochem 20:365–408
3. Borbas JE, Wheeler AP, Sikes CS (1991) Molluscan shell matrix phosphoproteins: Correlation of degree of phosphorylation to shell mineral microstructure and to in vitro regulation of mineralization. J Exp Zool 258:1–13
4. Veis A (1989) Biochemical studies of vertebrate tooth mineralization. In: Mann S, Webb J, Williams RJP (eds) Biomineralization: Chemical and biochemical perspectives. VCH, Weinheim, pp 189–222
5. Watabe N (1965) Studies on shell formation. XI. Crystal-matrix relationships in the inner layers of mollusk shells. J Ultrastruct Mol Struct Res 12:351–370
6. Travis DF, Francois CJ, Bonar LC, Glimcher MJ (1967) Comparative studies of the organic matrices of invertebrate mineralized tissues. J Ultrastruct Mol Struct Res 18:519–550
7. Bevelander G, Nakahara H (1980) Compartment and envelope formation in the process of biological mineralization. In: Omori M, Watabe N (eds) Tokai University Press, Tokyo, pp 19–27
8. Crenshaw MA (1972) The soluble matrix from *Mercenaria mercenaria* shell. Biomineralization 6:6–11
9. Weiner S (1979) Aspartic acid-rich proteins: Major components of the soluble organic matrix of mollusk shells. Calcif Tissue Int 29:163–167
10. Nakahara H, Mitsuo K, Bevelander G (1980) Five structure and amino acid composition of the organic "envelope" in the prismatic layer of some bivalve shells. Venus (Jpn J Malacol) 39:167–177
11. Iwata K (1975) Ultrastructure of the chonchiolin matrices in molluscan nacreous layer. J Fac Sci, Hokkaido Univ, IV 17:173–229
12. Wheeler AP, Rusenko KW, George JW, Sikes CS (1987) Evaluation of calcium binding by molluscan shell organic matrix and its relevance to biomineralization. Comp Biochem Physiol [B] 87:953–960

13. Weiner S (1983) Mollusk shell formation: Isolation of two organic matrix proteins associated with calcite deposition in the bivalve *Mytilus californianus*. Biochemistry 22:4139–4145
14. Keith JA, Stockwell SA, Fletcher MR, Napier K, Donnley TS, Kaplan DL (1991) Characterization of *Mytilus edulis* matrix proteins and the influence of temperature on crystallization of $CaCO_3$ under monolayers. Chem Eng News, May 11, 1991, p 72
15. Wheeler AP, Rusenko KW, Sikes CS (1988) Regulation of in vitro and in vivo $CaCO_3$ crystallization by fractions of oyster soluble matrix. Mar Biol 98:71–80
16. Rusenko KW, Donachy JE, Wheeler AP (1991) Purification and characterization of a shell matrix phosphoprotein from the American oyster. In: Sikes CS, Wheeler AP (eds) Surface reactive peptides and polymers. ACS, Washington, pp 108–124
17. Rusenko KW (1988) Studies on the structure and function of shell matrix proteins from the American oyster, *Crassostrea virginica*. PhD dissertation, Clemson University, Clemson, South Carolina
18. Wourms D, Weiner S (1986) Mollusk shell organic matrix: Fourier transform infrared study of the acidic macromolecules. J Exp Zool 237:11–20
19. Weiner S (1982) Separation of acidic proteins from mineralized tissues by reversed phase high-performance liquid chromatography. J Chromatogr 245:148–154
20. Butler WT, Bhown M, Dimuzio T, Linde A (1981) Noncollagenous proteins of dentin. Isolation and partial characterization of rat dentin proteins and proteoglycans using a three-step preparative method. Coll Res 1:187–199
21. Weiner S, Lowenstam HA (1977) Discrete molecular weight components of the organic matrices of mollusc shells. J Exp Mar Biol Ecol 30:45–51
22. Meenakshi VR, Hare PE, Wilbur KM (1971) Amino acids of the organic matrix of the gastropod shells. Comp Biochem Physiol [B] 40:1037–1043
23. Samata T, Sanguansri P, Cazaux C, Hamm M, Engles J, Krampitz G (1980) Biochemical studies on components of mollusc shells. In: Omori M, Watabe N (eds) The mechanisms of biomineralization in animals and plants. Tokai University Press, Tokyo, pp 37–47
24. Gordon J, Carriker MR (1980) Sclerotized protein in the shell matrix of a bivalve mollusc. Mar Biol 57:251–260
25. Gustafson GL, Milner LA (1980) Occurrence of N-acetylglucosamine-1-phosphate in proteinase I from *Dictyostelium discoideum*. J Biol Chem 255:7208–7210
26. Marsh ME (1986) Histidinoalanine, a naturally occurring cross-link derived from phosphoserine and histidine residues in mineral-binding phosphoproteins. Biochemistry 25:2392–2396
27. Kuboki Y, Fujisawa R, Tsuzaki M, Liu CF, Sasaki S (1984) Presence of lysinoalanine and histidinoalanine in bovine dentin phosphoprotein. Calcif Tissue Int 36:126–128
28. Addadi L, Moradian J, Shay E, Maroudas NG, Weiner S (1987) A chemical model for the cooperation of sulfates and carboxylates in calcite crystal nucleation: Relevance to biomineralization. Proc Natl Acad Sci USA 84:2732–2736
29. Simkiss K (1965) The organic matrix of the oyster shell. Comp Biochem Physiol 16:427–435
30. Weiner S, Hood L (1975) Soluble protein of the organic matrix of mollusc shells: A potential template for shell formation. Science 190:987–989
31. Akiyama M (1980) Thermal alteration experiments of scallop shells in relation to the diagenetic change of fossil proteins. In: Omori M, Watabe N (eds) The mechanisms of biomineralization in animals and plants. Tokai University Press, Tokyo, pp 257–262

32. Weiner S (1981) Repeating amino acid sequences in the soluble protein of mollusc shell organic matrices: Their involvement in crystal formation. In: Veis A (ed) The chemistry and biology of mineralized connective tissues. Elsevier, North Holland, pp 517–521
33. Donachy JE (1991) Atomic force microscopy and sequence analysis of a matrix protein from calcium carbonate biomineral. Chem Eng News, Mar. 11, 1991, p 72
34. Veis A (1991) Phosphoproteins as mediators of biomineralization. In: Sikes CS, Wheeler AP (eds) Surface reactive peptides and polymers. ACS, Washington, pp 1–12
35. Greenfield E (1987) In vitro mineral induction by soluble matrix from molluscan shells. PhD dissertation, University of North Carolina, Chapel Hill, North Carolina
36. Crenshaw MA, Linde A, Lussi A (1988) Biologically formed composites. In: Aksay IA, McVay GL, Stoebe TG, Wager JG (eds) Atomic and molecular processing of electronic and ceramic materials. Materials Research Society, Pittsburg, pp 99–107
37. Greenfield EM, Wilson DT, Crenshaw MA (1984) Ionotropic nucleation of calcium carbonate by molluscan matrix. Am Zool 24:925–932
38. Stetler-Stevenson WG, Veis A (1987) Bovine dentin phosphophoryn: Calcium ion binding properties of a high molecular weight preparation. Calcif Tissue Int 40:97–102
39. Lee SL, Veis A (1980) Cooperativity in calcium ion binding to repetitive, carboxylate-serylphosphate polypeptides and the relationship of this property to dentin mineralization. J Peptide Protein Res 16:231–240
40. Sikes CS, Wheeler AP (1988) Regulators of biomineralization. CHEMTECH 18:620–626
41. Sikes CS, Yeung ML, Wheeler AP (1991) Inhibition of calcium carbonate and calcium phosphate crystallization by peptides enriched in aspartic acid and phosphoserine. In: Sikes CS, Wheeler AP (eds) Surface reactive peptides and polymers. ACS, Washington, pp 50–71
42. Wheeler AP, Low KC, Sikes CS (1991) $CaCO_3$ crystal-binding properties of peptides and their influence on crystal growth. In: Sikes CS, Wheeler AP (eds) Surface reactive peptides and polymers. ACS, Washington, pp 72–84
43. Sikes CS, Wheeler AP (1986) The organic matrix of oyster shell as a regulator of calcification in vivo. Biol Bull 170:494–505
44. Watabe N (1981) Crystal growth of calcium carbonate in the invertebrates. Prog Crystal Growth Character 4:99–147
45. Addadi L, Weiner S (1985) Interactions between acidic proteins and crystals: Steriochemical requirements in biomineralization. Proc Natl Acad Sci USA 82: 4110–4114
46. Wheeler AP, Sikes CS (1984) Regulation of carbonate calcification by organic matrix. Am Zool 24:933–944
47. Berman A, Addadi L, Weiner S (1988) Interactions of sea urchin skeleton macromolecules with growing calcite crystals—a study of intracrystalline proteins. Nature 331:546–548

Index

SPRINGER NATURE

GPSR Compliance

The European Union's (EU) General Product Safety Regulation (GPSR) is a set of rules that requires consumer products to be safe and our obligations to ensure this.

If you have any concerns about our products, you can contact us on ProductSafety@springernature.com

In case Publisher is established outside the EU, the EU authorized representative is:

Springer Nature Customer Service Center GmbH
Europaplatz 3
69115 Heidelberg, Germany

FSC
www.fsc.org
MIX
Papier | Fördert
gute Waldnutzung
FSC® C083411

Kolibri 360 GmbH
Ferdinand-Jühlke-Straße 7
99095 Erfurt, Deutschland
produktsicherheit@kolibri360.de